Bryophytes - The State of Knowledge in a Changing World
http://dx.doi.org/10.5772/intechopen.102134
Edited by Jair Putzke

Contributors
Friedrich Altmann, David Stenitzer, María E. Holgado Rojas, Steven Paul Sylvester, Mitsy Diaz P.V. Sylvester, Marisol Saji Sayre, Alicia Kröpfl, Jair Putzke, Cassiane Furlan Lopes, Alice Lemos Costa, Marines De Avila Heberle, Ana Luiza Klotz Neves, Kamille Rodrigues Ferraz, Marina De Souza Falcão, Guilherme Henrique Mueller, Jorge Renato Pinheiro Velloso, Fernando Augusto Bertazzo Da Silva, Yuninda Riyana, Sasi Gendro Sari, Gunawan Gunawan

First published in London, United Kingdom, 2023 by IntechOpen
IntechOpen is the global imprint of INTECHOPEN LIMITED, registered in England and Wales, registration number: 11086078, 5 Princes Gate Court, London, SW7 2QJ, United Kingdom

British Library Cataloguing-in-Publication Data
A catalogue record for this book is available from the British Library

Additional hard and PDF copies can be obtained from orders@intechopen.com

Bryophytes - The State of Knowledge in a Changing World
Edited by Jair Putzke
p. cm.
Print ISBN 978-1-80356-938-3
Online ISBN 978-1-80356-939-0
eBook (PDF) ISBN 978-1-80356-940-6

Meet the editor

Jair Putzke obtained a degree in Biological Sciences from the Universidade de Santa Cruz, Brazil, in 1990, and a master's degree in Cryptogams and a Ph.D. in Botany from the Universidade Federal de Pernambuco, Brazil, in 1994 and 2003, respectively. He is currently a professor at the Universidade Federal do Pampa Campus São Gabriel, Brazil. He has experience in botany with an emphasis on mycology, working mainly on fungi/bryophytes relationships, mycology, lichens, taxonomy, myxomycetes, botany, ecology, teaching, and environmental education. He has published 120 scientific articles, 23 books, and 13 book chapters. Dr. Putzke has studied bryology for more than 30 years, mainly in Antarctica.

Contents

Preface

The world is changing rapidly due to anthropic interference, and groups of living beings little known by the population, both in their culture and in practical applications, are the most impacted. This impact can have very big implications if we consider the fragility of the different species of Bryophyta. Bryophytes are the most primitive group of plants, generally consisting of leaves one-cell thick, stems without conductive tissues, and sexual reproduction by male gamete mobile by flagella and ultimately resulting in the formation of spores asexually.

The group is divided into mosses (Phylum Bryophyta), hornworts (Anthocerotophyta), and liverworts (Marchantiophyta), with differences in the thallus and the level of cellular organization between them. Stomata can already be found (rarely), but they are still primitive and not very necessary judging from the one-cell thickness of the leaf of most species. In this way, they are very simple and fragile and can be quickly eliminated from an area by some atmospheric pollutant (since they basically feed on what is dissolved in the relative humidity of the air) or by forest fires.

If we consider the data on the impact of the great forest fires occurring in several important ecosystems around the globe, the impact on Bryophyta is gigantic. Despite this, few studies on Bryophyta have been conducted simply because there are not enough scientists.

Attracting young people's attention to this group of plants so that they start to study them or at least protect them depends on the existence of books and other literature to pave the way for future research. This book is an attempt to further our understanding and knowledge of Bryophyta. It is organized into 4 sections and 6 chapters.

Chapter 1 is an introductory chapter that discusses different areas where the Bryophyta have been subjects of applied research and what can be done to train and attract more scientists to study this group. The Bryophyta are presented as an extremely resistant group to different environmental conditions, even though they are very fragile. The possibility of being used as bioindicators, as accumulators of pollutants, or even as carbon reserves at a time of great need to reduce their rates in the atmosphere has led to considering the group vitally important in controlling the agents that are causing climate change and harmful effects on human health.

Chapter 2, "Diversity of Epiphytic Mosses (Bryophyta) in Forests of *Polylepis* (Rosaceae) in the Urubamba Mountain Range, Cusco, Peru", presents results of a taxonomic survey of Bryophyta in this understudied region. The work reveals three new occurrences for the country and fourteen new occurrences for the Cusco region (4300–4800 m of altitude). The study reveals how important it is to preserve the formations with this bush to guarantee the maintenance of the diversity of mosses in the region. Similar studies in similar formations in other high-altitude regions may reveal many novelties in terms of phytogeography of different species.

Chapter 3, "Bryophyta around Syamsudin Noor International Airport, South Kalimantan, Indonesia", presents the results of a study of diversity in the area around an airport in Indonesia. The study reveals interesting data that may allow for similar studies at other airports around the world, revealing the impact of air transport on moss communities. The data may also allow monitoring changes in the community in the future with the increase in the number of flights and the increase in the population, which may indicate that the species are resistant to pollution and can be used as bioindicators of air quality.

Section 3, "Interactions: Bryophytes and Other Organisms", begins with Chapter 4, "Bryophilous Agaricomycetes (Fungi, Basidiomycota): A Review to Brazil". This chapter reviews the mushroom species associated with bryophytes, which are considered a type of parasite. Few studies have been carried out on organisms associated with bryophytes, especially cyanobacteria, algae, fungi, and lichens, whose ecological relationship still needs to be better explained. In a country with a vast territory like Brazil, a more in-depth survey of these groups still does not exist. Considering the worldwide surveys, more studies are also needed, especially with the combined efforts of specialists in these areas. In Brazil, most species of Agaricomycetes fungi parasitizing mosses have been found at high altitudes and with high humidity.

Chapter 5, "Why Are Moss Biocrusts Necessary for System Conservation in a Semiarid Region of Southern Argentina?", evaluates moss biocrusts, establishing their composition in mosses as the dominant one and verifying the impact that inadequate management, fire, and other anthropic activities determine in this almost desert formation. They are associated with the shrub steppe and herbaceous formations, having in their composition cyanobacteria, algae, fungi, and lichens. Their fragility in extreme conditions requires the creation of more conservation areas for this rich, complex species and adequate management in areas with agriculture and livestock.

Finally, in the last section, Chapter 6 "Protein Glycosylation in Bryophytes Differs Subtly from That in Vascular Plants", examines the biochemical novelties found in bryophytes and how they differ from other groups of plants, especially with regard to N-glycan biosynthesis. The work demonstrates the rich world of Bryophyta biochemistry, which could guarantee new drugs in the future.

Special thanks go to Author Service Manager Maja Bozicevic at IntechOpen for his outstanding work throughout the preparation and publication of this book.

Special Sincere thanks also go to the chapter authors who answered a call in the middle of the end of a pandemic and dedicated time in their research to contribute to the publication of this book.

Jair Putzke
Universidade Federal do Pampa,
Curso de Pós graduação em Ciências Biológicas,
São Gabriel, Brazil

Section 1

Introduction

Chapter 1

Introductory Chapter: Bryophytes - Why Should We Study Them?

Jair Putzke

1. Introduction

Bryophytes are indeed an amazing group! For more than 470 million years, they have advanced over terrestrial environments, being responsible for the formation of the first layers of organic soil. This allowed the installation of other terrestrial plants and was responsible for about 1/3 of the oxygen available for the breathing of all living beings that later occupied these environments [1].

They cover a very large area of the planet, generally where other plants cannot grow. If we consider mountains, much of the tundra and taiga and the bark of trees which are covered with mosses, the photosynthetic area is probably greater than that of other plants combined, even though they are so small and delicate. In boreal forests, mosses may be the smallest components in biomass, but they represent the greatest coverage and have the highest primary productivity, in addition to being faster in the accumulation of nutrients, contributing to the thermal development of the environment with consequent development of the permafrost. In this way, they positively affect nutrient cycling in these ecosystems, affecting also the vascular plants [2]

Bryophytes are of unparalleled resistance, sprouting again after hundreds of years covered by ice (from 400 to 1530 years—already detected) or even if kept for years in herbaria in conditions of extreme dehydration and darkness. There are even specific genes to resist to climate changes [3–6].

2. Bryophytes and the environmental changes

Climate change is affecting the group very deeply, in particular because they are very dependent on water, including for their sexual reproduction, as they have flagellated antherozoids. Even in environments where they are the dominant flora, such as in maritime Antarctica, species may have their occurrence threatened in the face of climate change. This is also the case of boreal areas and high mountains ecosystems [7]. Bryophytes can even be used for monitoring and as an indication of the effects of climate change on living organisms. They have even been indicated as "canaries in the coal mine" with regard to their behavior in relation to the external environment [8]. As a result, they are symbols of eternity in some cultures.

This group has very simple thalli, with leaves generally formed by only one layer of cells, stems without conductive tissues and rhizoids only one cell thick, which obtain their nourishment directly from atmospheric humidity, from the soil or substrate where they are fixed. Many react quickly in the presence of pollutants and can be used to monitor and control them. Morphological and even genetic modifications in relation to the toxicity of polluting metals or even the stress caused by their presence can be followed, making them first-class bioindicators [9, 10]. Many species are extremely sensitive to pollutants, suffering some injuries or even resulting in the death of the entire thallus, while others are very resistant, even managing to retain pollutants, preventing them from affecting other species in the environment [11]. The reduction of pollutants in large urban centers is already being done using bryophytes in systems that filter and purify the atmospheric air with great efficiency.

There are even species that live submerged in water and that can collaborate in the removal of pollutants from these systems without suffering damage, as is the case of *Leptodyction riparium*, *Fontinalis antipyretica,* and *Scapania undulata*, among others [12, 13]. Some aquatic species are even reported as accumulating very rare chemical elements in nature such as Nd, Gd, Ho, Er, Tm, Lu, La, Ce, Sm, Eu, Tb, and Dy [14].

3. More bryologists for more knowledge and preservation

The number of specialists dedicated to the study of the group, from the most basic to the applied sciences, is very small. Finding more young people who are interested in the group is now a fundamental task in all countries, as we are losing habitat and species due to anthropic interference. In this way, this group will only be known, appreciated, and preserved if more specialist researchers will be formed. As it is one of the least studied groups, much still needs to be discovered, starting with the basic taxonomy and distribution of species, their relationships with other groups, especially with vascular plants and fungi, among others. As an example of this last topic, it can be mentioned that almost nothing is known about their pathogens and how they are relating to mosses in the face of climate changes [15, 16].

Studies for the conservation of bryophytes are also very limited, requiring work with in vitro cultivation, tissue culture, with the maintenance of spore banks, creation of lists of endangered species and location of endangered populations and maintenance cultures of the genetic diversity within species [17].

Bryophyte extracts were tested against many groups of insects as to control their populations. Test against lepidopterans, dipterans and/or coleopterans of some species have had good results for antifeedant, insecticide, or as developmental inhibitor [18].

Bryophytes are generally very fragile and delicate, offering no barriers to pathogen attack. Instead, they are rich in secondary metabolites that guarantee a strong defense not only against pathogenic agents but also against herbivores. These metabolites have wide application in medicine and pharmacy, in addition to biotechnology. They have long been used in medicine and are increasingly associated with antibacterial, antiviral, and antifungal properties. Others have been tested for their antitumor activities. Many also confer special aromas to some species, also having application in this area [19, 20].

The practical application of bryophytes in environmental monitoring, in the control and reduction of pollutants, in landscaping, and in energy generation, for example, needs to be more publicized to attract the attention of other areas to the theme. Studies in electricity generation and other practical uses are also needed to reinforce their preservation [16, 21].

In this way, gathering in a book additional information about bryophytes can attract more people to study the group and to practice their conservation. One always has in mind the creation of conservation units based mainly on arboreal components or mammals, and bryophytes always indirectly compete in these projects, but without special attention. It is necessary to organize conservation areas with a focus also on threatened bryophytes.

In vitro cultivation banks that include all species already catalogued are also needed. The species known number is so small (compared with other groups), and the area occupied by massive cultivation would also be so reduced that a worldwide effort could create an in vitro germplasm bank of all redeemable species, starting with those deposited in herbaria. The effort could start with the in vitro maintenance of newly described species, as for example. This would mean a worldwide effort in this direction, but the group still lacks enthusiasm and funding in this regard.

Even though they are increasingly associated with all these areas of knowledge (and there are many others), this is perhaps the group in which less research is being done in all areas. The group was the first to massively occupy the terrestrial environment and has since evolved, longer than any other group of photosynthetic organisms, a unique set of chemicals to survive all the biotic and abiotic stresses they have been subjected to. They are also resistance heroes for that reason and deserve a prominent place in science, something we are still far from.

Even with all these known applications, why don't you have moss gardens spread as a healthy practice among the inhabitants of the entire planet? Why we do not have a garden of medicinal bryophytes? Why don't we have bryophyte filters to improve the air in our living and working environments? Much of this is answered with: because no one knows all this!

Attracting interest of new enthusiasts to this spectacular group is the focus of this book. Good reading.

Acknowledgements

To the Universidade Federal do Pampa campus São Gabriel in special for the efforts of our ex-director Chariston Andre Dal Belo and to the director Luciana Borba Benetti regarding our laboratory facilities.

Author details

Jair Putzke
Programa de Pós-Graduação em Ciências Biológicas, Universidade Federal do Pampa, São Gabriel, Brazil

*Address all correspondence to: jairputzke@yahoo.com.br

References

[1] Beike AK, Lang D, Zimmer AD, Wüst F, Trautmann D, Wiedemann G, et al. Insights from the cold transcriptome of Physcomitrella patens: Global specialization pattern of conserved transcriptional regulators and identification of orphan genes involved in cold acclimation. The New Phytologist. 2014;**205**(2):869-881. DOI: 10.1111/nph.13004

[2] Lenton TM, Dahl T, Daines SJ, Mills BJW, Ozaki K, Saltzman MR, et al. Earliest land plants created modern levels of atmospheric oxygen. Proceedings. National Academy of Sciences. United States of America. 2016;**113**:9704-9709

[3] La Farge C, Williams KH, England JH. Regeneration of little ice age bryophytes emerging from a polar glacier with implications of totipotency in extreme environments. Proceedings of the National Academy of Sciences, USA. 2013;**110**:9839-9844

[4] Lang D, Weiche B, Timmerhaus G, Richardt S, Riaño-Pachón DM, Corrêa LGG, et al. Genome-wide phylogenetic comparative analysis of plant transcriptional regulation: A timeline of loss, gain, expansion, and correlation with complexity. Genome Biology and Evolution. 2010;**2**:488-503

[5] Sabovljević MS, Ćosić MV, Jadranin BZ, Pantović JP, Giba ZS, Vujičić MM, et al. The conservation physiology of bryophytes. Plants (Basel). 2022;**11**(10):1282. DOI: 10.3390/plants11101282

[6] Vujičić M, Sabovljević A, Šinžar-Sekulić J, Skorić M, Sabovljević M. In vitro development of the rare and endangered moss Molendoa hornschuchiana (Hook.) Lindb. ex Limpr. (Pottiaceae, Bryophyta). HortScience. 2012;**47**:84

[7] Oechel WC, Van Cleve K. The role of bryophytes in nutrient cycling in the Taiga. In: Forest Ecosystems in the Alaskan Taiga, Ecological Studies 57 Van Cleve FS, Chapin PW, Flanagan LA, Dyrness CT, pp. 121-137, Springer, New York. 1986.

[8] Bramley-Alves J, King DH, Robinson SA. Dominating the Antarctic Environment: Bryophytes in a Time of Change. In: Hanson DT, Rice SK, editors. Photosynthesis in Bryophytes and Early Land Plants. Dordrecht: Springer; 2014. pp. 309-324

[9] Hohenwallner D, Zechmeister HG, Moser D, Pauli H, Gottfried M, Reiter K, et al. Alpine bryophytes as indicators for climate change: A case study from the Austrian Alps. In: Tuba Z, Slack NG, Stark LR, editors. Bryophyte Ecology and Climate Change. Cambridge, UK: Cambridge University Press; 2011. pp. 237-250

[10] Stanković JD, Sabovljević AD, Sabovljević MS. Bryophytes and heavy metals: A review. Acta Botanica Croatica. 2018;**77**:109-118

[11] Rühling A, Tyler G. Changes in the atmospheric deposition of minor and rare elements between 1975 and 2000 in south Sweden, as measured by moss analysis. Environmental Pollution. 2004;**131**:417-423

[12] Pratas J, Favas PJ, Varun M, D'Souza R, Paul MS. Distribution of rare earth elements, thorium and uranium in streams and aquatic mosses of Central Portugal. Environment and Earth Science. 2017;**76**(4):156

[13] Esposito S, Loppi S, Monaci F, Paoli L, Vannini A, Sorbo S, et al. In-field and in-vitro study of the moss Leptodictyum riparium as bioindicator of toxic metal pollution in the aquatic environment: Ultrastructural damage, oxidative stress and HSP70 induction. PLoS One. 2018;**13**(4):e0195717. DOI: 10.1371/journal.pone.0195717

[14] Govindapyari H, Leleeka M, Nivedita M, Uniyal PL. Bryophytes: Indicators and monitoring agents of pollution. NeBIO. 2010;**1**:35-41

[15] Carey C. How physiological method sand concepts can be useful in conservation biology. Integrative and Comparative Biology. 2005;**45**:4-11. DOI: 10.1093/icb/45.1.4

[16] Roads E, Longton RE, Convey P. Millennial timescale regeneration in a moss from Antarctica. Current Biology. 2014;**24**:R222-R223. DOI: 10.1016/j.cub.2014.01.053

[17] Sinha S, Singh A, Sinha D, Chatterjee R. A review on bryophytes as key bio-indicators to monitor heavy metals in the atmosphere. International Journal of Plant and Environment. 2021;**7**(1):49-62

[18] Rowntree JK, Pressel S, Ramsay MM, Sabovljevic A, Sabovljevic M. In vitro conservation of European bryophytes. In Vitro Cell. 2011;**47**:55-64

[19] Alves RJM, Miranda TG, Tavares-Martins ACC. Scientometric approach to bryophytes bioactivity: The antiinsects potential and the perspectives for the XXI century. Research, Society and Development. 2020;**9**(12):475

[20] Dziwak M, Wróblewska K, Szumny A, Galek R. Modern use of bryophytes as a source of secondary metabolites. Agronomy. 2022;**12**:1456

[21] Putzke J, Ferrari FR, Schaefer CEGR. Discovery of a large population of Hygrolembidium isophyllum (Lepidoziaceae, Marchantiophyta) in the South Shetland Islands, Antarctica. Polar Research. 2020;**39**:1-4

Section 2

Diversity

Chapter 2

Diversity of Epiphytic Mosses (Bryophyta) in Forests of *Polylepis* (Rosaceae) in the Urubamba Mountain Range, Cusco, Peru

Marisol Saji Saire, María E. Holgado Rojas,
Mitsy Diaz P.V. Sylvester and Steven Paul Sylvester

Abstract

The diversity of mosses was evaluated in four *Polylepis* (Rosaceae) forests in the Cordillera del Urubamba, Cusco region – Peru. Epiphytic mosses were collected from the lower base of the trunk, canopy area, and terminal branches in paper bags, selecting specimens with sporophytes. Twenty-seven species distributed in seven orders, 13 families, and 17 genera were determined. Three species are new records for Peru, that is, *Neckera ehrenbergii*, *Zygodon quitensis,* and *Didymodon challaense*. Fourteen species are reported for the first time in the Cusco region. These chorological novelties demonstrate the importance of *Polylepis* forests as stores of cryptogamic diversity.

Keywords: diversity, mosses, canopy, terminal branches, new records

1. Introduction

In the upper parts of the Cusco region – Peru, the vegetation is mostly in the form of grasslands and small shrubs, with the genus *Polylepis* (Rosaceae) being the only dominant tree element. These forest patches are vital for protecting the soil against erosion, retain nutrients, sediment, and produce a vital amount of oxygen, and host a unique diversity of birds and plants, among other organisms. Despite their importance, these forests are in danger of extinction due to the growing pressure of the Andean population, as a product of numerous economic, social, and cultural factors [1]. There are several studies carried out on *Polylepis* in terms of flora and fauna in Peru and especially in Cusco, but not so on the epiphytic mosses that are found predominantly in these arboretums.

The *Polylepis* forests in the Urubamba valley constitute one of the largest population concentrations and the largest center of genetic diversity in the Andes with six species of the 10 cited for the national territory, *Polylepis besseri, P. racemosa, P. subsericans, P. pauta, P. sericea,* and *P. microphylla*, the same ones that house a great diversity of mosses that play a fundamental role in the conservation and maintenance of ecosystems due to their hydrophilicity, responsible for keeping these habitats moist; making necessary to know its diversity.

The Musci are the most complex group within the Bryophytes, not only because of the morphology of their gametophytes but particularly because of the structure of their sporophytes. They make up the group of Bryophytes with the most species; it is estimated that worldwide there are between 8000 and 12,800 species in 900 genera. The number of mosses estimated for the Andean region of Colombia, Ecuador, and Peru is likely to reach around 900–950 species. Among these countries, Peru requires more work both in terms of additional inventories and published results; however, around 829 taxa are currently known for the entire territory, of which 797 are reported for the Andean region [2]. There is a gradual increase in the diversity of taxa according to elevation, with a maximum between 2500 and 3000 m, beyond this altitude interval, there is a marked decrease in diversity, notably above 3500 m. Thus, in Bolivia, a greater diversity was recorded between 2000 and 2500 m, followed by the 2500–3000 m interval [3].

In this regard, Menzel [4] cites 181 species for the department of Ayacucho, the vast majority collected by Hegewald & Hegewald, who carried out explorations in various departments of Peru between 1973 and 1977, including Ayacucho, some were carried out in the province of Huamanga. Opisso [5, 6] carried out an inventory of pleurocarpous mosses in the province of San Ignacio – Cajamarca, registering 40 species, 32 genera, and 15 families. He also recorded 55 species of mosses for the Pomahuaca-Cajamarca district with 47 genera and 30 families. Opisso and Churchill [7] carried out a study in the surroundings of the Yanachaga-Chemillén National Park in the department of Pasco, in which they indicate the existence of 134 species of bryophytes distributed among 92 genera and 45 families (32 liverworts, 2 hornworts, and 100 mosses).

In the Cusco region, studies on Bryophytes began in 1921 with Fortunato L. Herrera in his work "Flora del Departamento del Cuzco" reporting 26 species of mosses, among which he mentioned *Hygrodicranum herrerai* Williams, *Leptodontium brachyphyllum*, *Tortula affinis* Hampe, *Bryum andicola* Hook, *Ortothrichum elongatum* Taylor, and *Polytrichum antillarum* Rich. Galiano [8] carried out a study on the flora of the high tropical Andean Forest of Yanacocha, in the province of Urubamba, registering 13 species of mosses, belonging to 10 genera and eight families. Tupayachi [9], when evaluating the flora of the Cordillera del Vilcanota, reported 35 species of bryophytes, belonging to 24 genera and 14 families. Galiano et al. [10], when studying the flora of the Cusco Valley, determined 14 species of mosses. Acurio [11] evaluated the diversity and distribution of mosses in the area of Wiñay Wayna, Historic Sanctuary of Machu Picchu-Cusco in an altitude range of 2500–3100 m, finding 129 species in 72 genera and 29 families. Subsequently, Huallparimachi et al. [12] reported for the Sanctuary 50 species of mosses included in 23 genera and 15 families, highlighting *Bartramia*, *Brachythecium*, *Bryum*, *Campylopus*, *Lepyrodon*, *Neckera*, *Plagiothecium*, *Polytrichum*, *Sematophyllum*, *Sphagnum*, *Tortula*, *Trichostomun*, *Thuidium*, and *Zygodon*.

For its part, Carhuapoma [13] carried out the study of mosses of the Historical Sanctuary of the Pampa de Ayacucho in an altitude range of 3350 to 4100 m, registering 67 species belonging to 45 genera, 21 families, and 11 orders. The Pottiaceae family was the most representative with 10 species, followed by Orthotrichaceae, Brachytheciaceae, and Bartramiaceae with six species each, reporting 23 new records for Ayacucho and the third report for Peru of *Philonotis scabrifolia* (Hook. f. & Wilson) Braithw. expanding its distribution to the south of the country. Despite all these reports, there is still little progress in this division, and it is necessary to encourage these studies and expand our knowledge about the existing bryoflora in Peru and particularly in the Cusco region. The present study is carried out in order to determine the composition of mosses in each of the *Polylepis* forests in the Urubamba Mountain range.

2. Methods

3. Study area

The scope of study territorially includes the Huarán basin with the towns of Canchacancha (13° 14′ 34.9" S and 72° 01′ 13.4" W) located at 4548 m.a.s.l. and Contorkayku (13° 16′ 03.7" S and 72° 01′ 02" W) at 4310 m.a.s.l. between the limits of the provinces of Calca and Urubamba; the Wayoqhari-Yanacocha basin, locality K′elloq'ocha (13° 16′ 34.3" S and 72° 03′09.7" W) at 4343 m.a.s.l. belonging to the district of Huayllabamba, and the Manthanay basin, locality Manthanay (13° 12′ 08.4" S and 72° 08′ 42.8" W) at 4778 m.a.s.l. located in the heights of the Yanahuara Valley in the province of Urubamba, Peru.

3.1 Natural life zones

3.1.1 Subtropical sub-Andean very humid páramo (pmh-SaS)

This life zone also known as wet puna in the South of Peru; it presents a total annual precipitation that varies between 640 mm and 800 mm and an average annual biotemperature between 6°C and 3°C. It is a cold climate zone. According to the Holdridge Diagram, this life zone has potential evapotranspiration that varies between a quarter (0.25) and a half (0.5) of the average total precipitation per year. Life includes the dense forests of *Polylepis*, located from 3900 to 4500 m. of altitude, with a thick forest of shrubs and herbs as well as an abundant presence of mossy mattresses between the rocks of the forest floor and the trunks of the trees, which is due to the high humidity prevailing inside [14].

3.1.2 Subtropical sub-Andean pluvial tundra (tp-SaS)

Life zone includes the ecological system of the cold Andean desert of the humid puna. It is located on the very humid-subalpine and subtropical paramo and below the Nival floor. It has a cold climate, the soils are mostly rocky due to erosion of the old glacial mountains, in whose cracks there is scattered and discontinuous vegetation conditioned and adapted to the daily alternation of ice-thaw and where *Polylepis subsericans* ascends through the tongues of cryoturbated soils in a clear process of colonization toward the rocky crests caused by glacial erosion. Altitudinally, they are above 4500 m [14].

4. Sampling

Using the protocol of Gradstein et al. [15], four forests were selected, in which 16 *Polylepis* arboretums were evaluated. A total of 10 plots per arboretum were evaluated (four in the bolus, four in the canopy, and two in the terminal branches). Each plot was 20 × 30 cm in the bolus and canopy, and in branches the plots were 10 × 60, making a total of 600 cm^2 each plot.

In each forest, epiphytic mosses were collected from the lower base of the trunk, canopy area, and terminal branches. The collection was made using craft paper bags and/or recycled paper envelopes, preferably selecting specimens with a sporophyte as this is often necessary for identification.

4.1 Identification process

To identify the species in the laboratory, it was necessary to rehydrate the samples, so that they recover their natural form. It was enough to submerge a few plants in water for them to rehydrate in a few seconds. In some species of phyllidia or thicker thallus, this operation took a little longer, but it is almost never more than a minute.

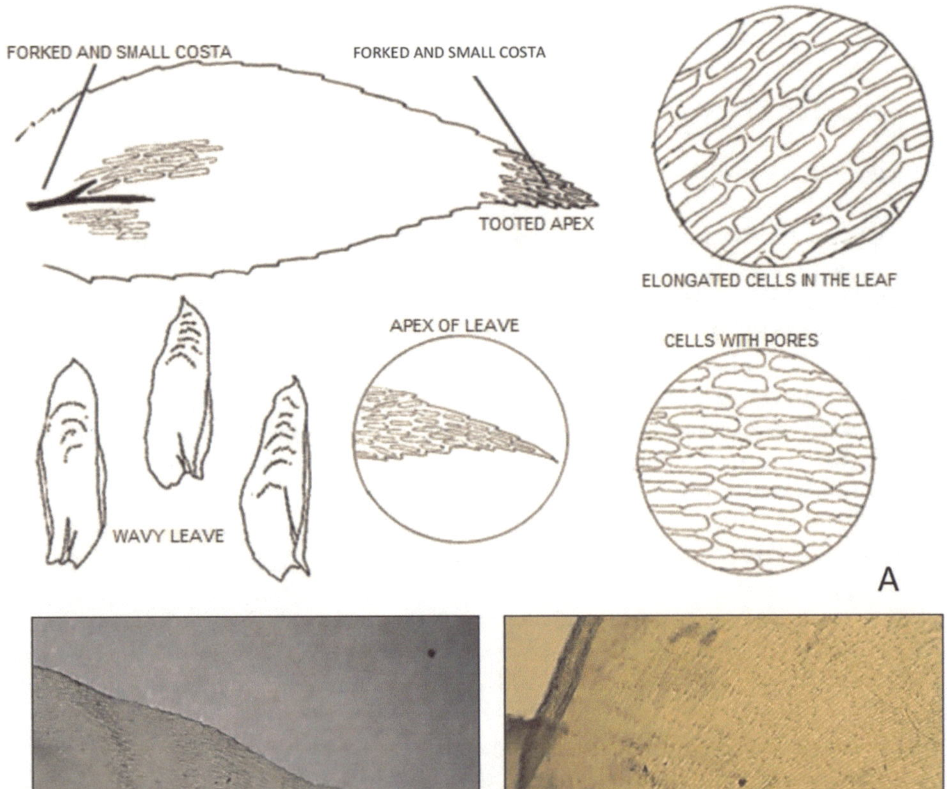

Figure 1.
A. Morphological characteristics of Neckera ehrenbergii *Müll. Hal.; B. undulate leaf 40x; C. bifurcate costa at leaf base 40x.*

Once the sample was rehydrated, with the help of fine-tipped tweezers and a pair of lancets, the parts that we are interested in observing were separated: phyllidia, caulydia, capsules, etc. Something important to keep in mind, before separating the samples, is that the details of the plant's shape must be observed, such as the arrangement of the phyllidia on the plant (distichous, more or less together or separated) and the type of branching.

Cross sections of the phyllidia of many species of mosses were also made to observe the structure of the central nerve, the thickness (number of cells) of the phyllidia, the presence of papillae or nipples, etc. (**Figures 1–8**).

The main characteristics observed were as follows:

Habit (acrocarpic, pleurocarpous moss).

Shape of the phyllidia.

Shape and dimensions of the cells of the phyllidia, presence of papillae, etc.

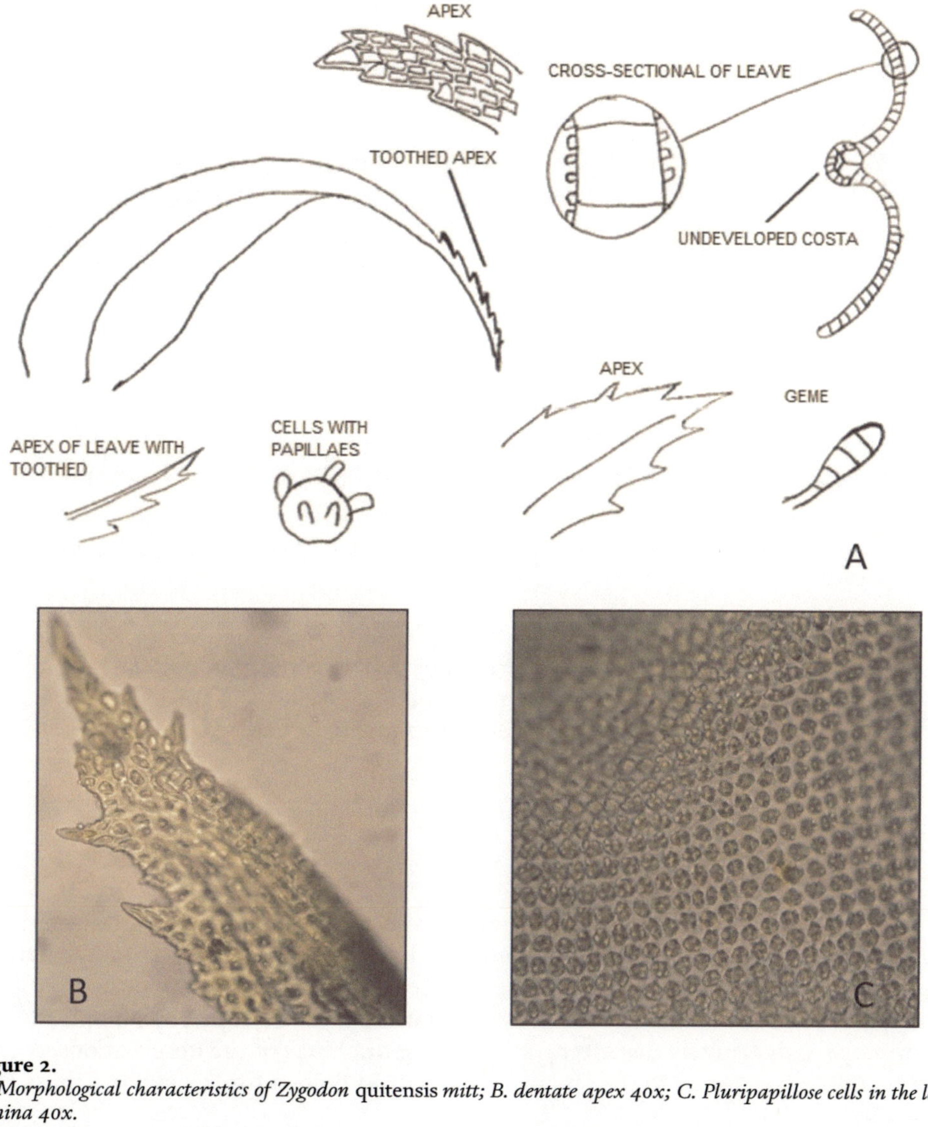

Figure 2.
A. Morphological characteristics of Zygodon quitensis *mitt; B. dentate apex 40x; C. Pluripapillose cells in the leaf lamina 40x.*

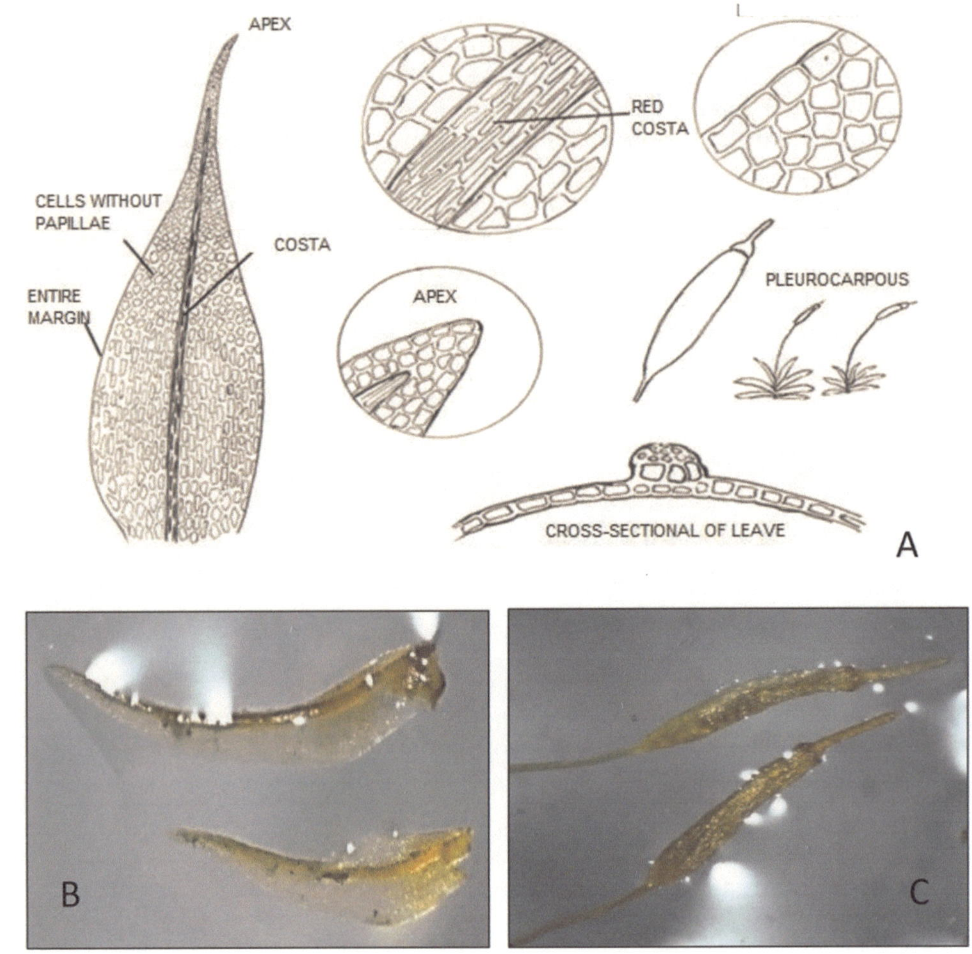

Figure 3.
A. Morphological characteristics of Didymodon challaense *(broth.) R.H. Zander; B. leaves 10x; C. Acrocarppic sporophytes 20x.*

Characteristics of the central nerve of the phyilidia.
Shape and dimensions of the cells of the phyllidia, presence of papillae, etc.
Characteristics of the margin of the phyllidia.
Presence of gems or other vegetatively reproducing structures.
Characteristics of the sporophyte.

5. Taxonomic identification

To determine the species, we first proceeded to separate the samples collected from other nonvascular epiphytes (lichens and liverworts), as well as from the bark of the trees, then proceeded to identify the samples following dichotomous keys [16], photographs of each of the species were taken using a stereoscope and an optical microscope, exhaustively describing and making drawings of the observations. Some of the samples were sent to the city of Santa Cruz in Bolivia for review by a moss specialist.

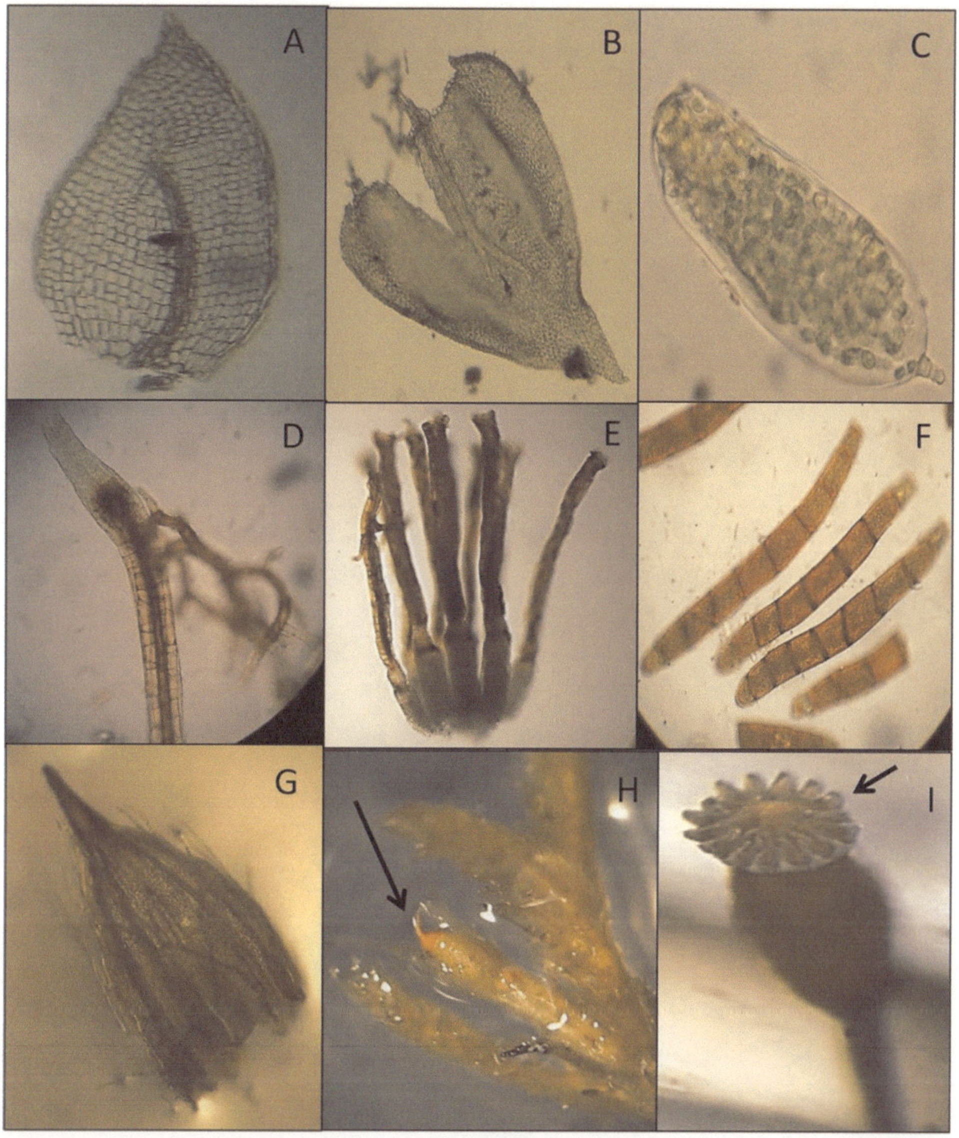

Figure 4.
A. Simple costae of leaf of Daltonia *sp.; B. forked costae of leaf in* T.peruvianun *mitt. 40x; C. leaf gemmae of* Zygodon *sp. 100x; D. Archaegonium of* Z. fragilis *H. rob. 100x; E. Archaegonia of* O. elongatum *Taylor 100x; F. propagules of* Zygodon quitensis *mitt. 100x; G. Calyptrae of* Semathophllum swartzii *(Schawägr) W.H. Welch & H. a, H. Crum; H. Operculum of* Neckera ehrenbergii *Müll. Hal. I. Peristomium of* Bartramia *sp.*

6. Results

In the forests of Manthanay K'elloq'ocha Contorkayku and Canchacancha, a total of 27 morphospecies of moss were found, which are distributed in seven orders, 13 families, and 17 genera. Twenty of these mosses are identified to species, six to genus, and one only at the family level (**Table 1**).

Three species are new records for Peru, that is, *Neckera ehrenbergii* Müll. Hal., *Zygodon quitensis* Mitt., and *Didymodon challaense* (Broth.) R.H. Zander. Likewise, 14 species are reported for the first time for the Cusco region, that is, *Bartramia potosica*

Figure 5.
Gametophytes and sporophytes of **a**. Orthotrichum elongatum *Taylor; B.* Zygodon reinwardtii *(Hornsch.) a. Braun; C.*: Lepyrodon tomentosus *(hook.) mitt.; D.* Neckera ehrenbergii *Müll. Hal.; E.* Daltonia trachyodonta *mitt.; F.* Daltonia *sp. costa.*

Mont, *Brachythecium occidentale* (Hampe) A. Jaeger, *B. andicola* Hook, *Bryum subapiculatum* Hampe, *Daltonia trachyodonta* Mitt, *Campylopus areodictyon* (Müll. Hal.) Mitt, *Chorisodontium mittenii* (Müll. Hal.) Broth, *Braunia cirrhifolia* (Mitt.) A. Jaeger, *Neckera andina* Mitt, *Orthotrichum elongatum* Taylor, *Zygodon fragilis* H. Rob, *Leptodontium tricolor* (R.S. Williams) R.H. Zander, *Syntrichia andicola* (Mont.) Ochyra and *Lepyrodon tomentosus* (Hook.) Mitt.

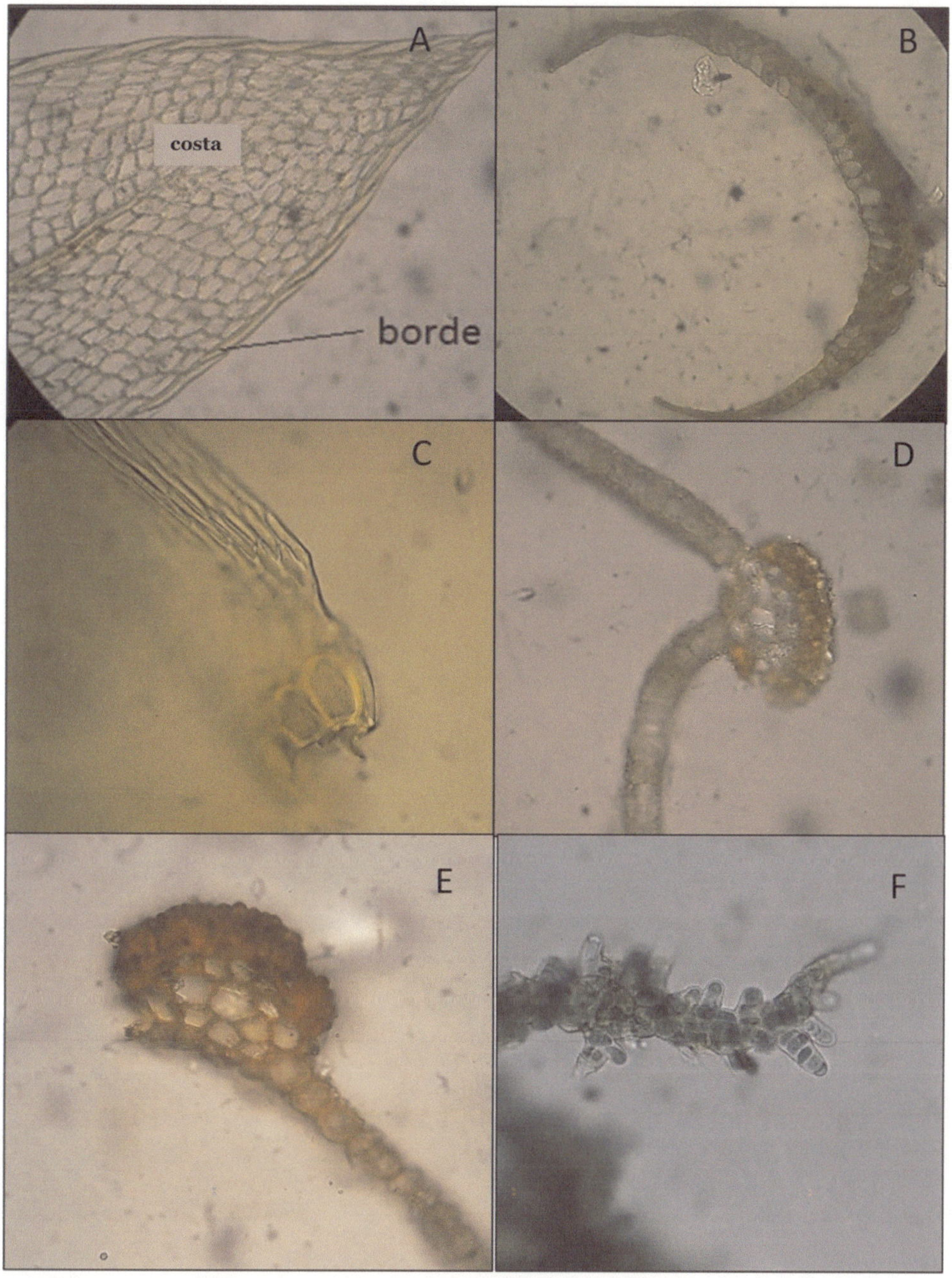

Figure 6.
A. Leave with costa of Daltonia *sp.*, 100x; *B. homogeneous leave cross-sectional in* Chorisodontium mittenii (Müll. Hall.); *C. alar cells in* Sematophyllum swartzii *(Schwägr.) W.H. Welch & H.a. 40x; D. Heterogeneous leave cross-sectional in* Daltonia trachyodonta mitt.; *E. Heterogeneous leave cross-sectional in* Syntrichia andicola *(Mont.); F. leave geme in* Braunia cirrhifolia *(mitt.) a. jaeger. 100x.*

As seen in **Table 1**, some morphospecies are almost exclusive to a single forest, such as *Orthotrichum sp.*, *B. subapiculatum* Hampe, and *Didymodon challaense* (Broth.) R.H. Zander. for the Canchacancha forest, *B. potosica* Mont. *Bartramia sp.* for Kelloq'ocha and *Syntrichia sp.* for Mantanay. Other species are more common, for

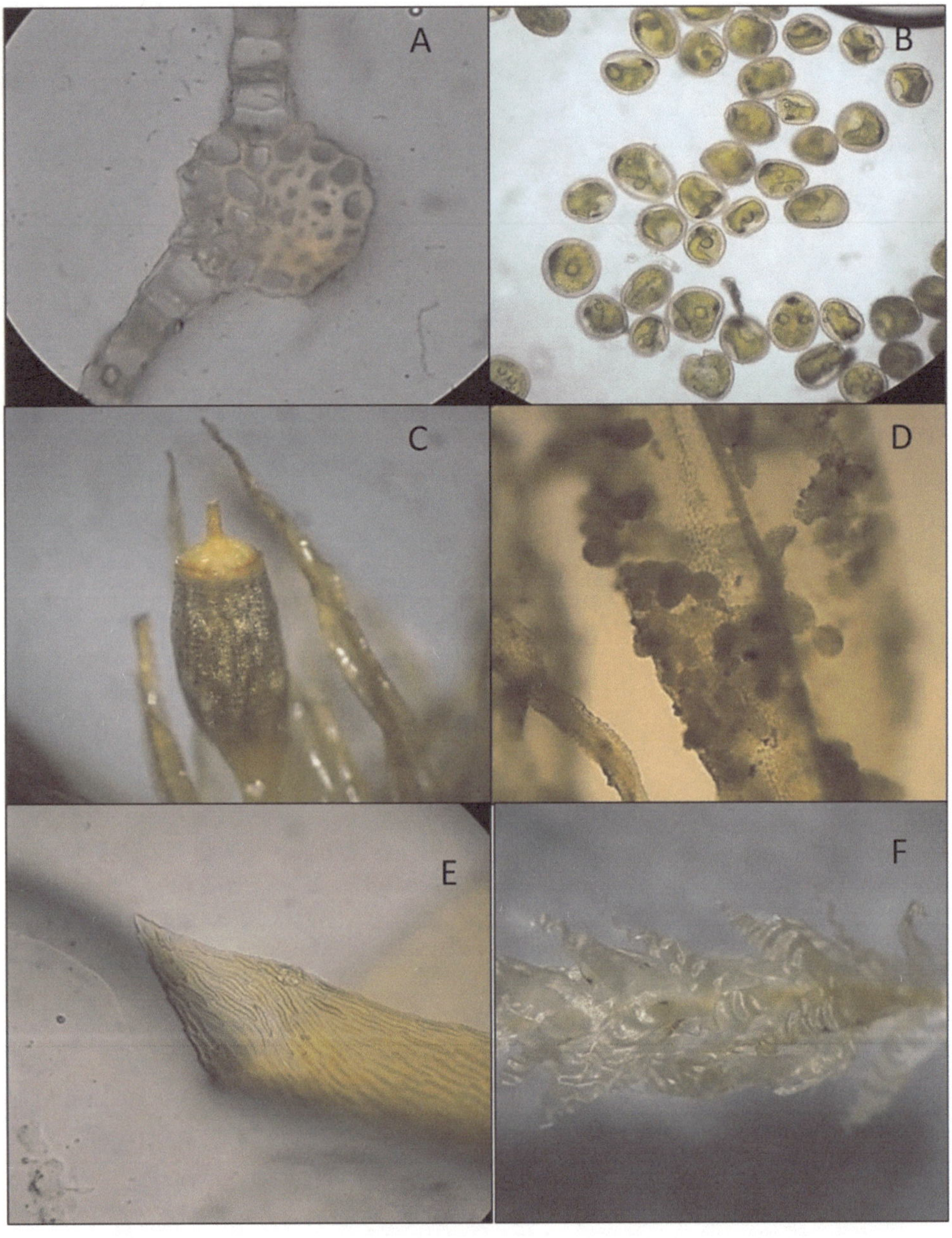

Figure 7.
A. Heterogeneous leave cross-sectional Bryum andicola *hook. 40x B. Gemes,* Orthotrichum elongatum *Taylor.* 100x; *C. sporophyte with operculum in* Orthotrichum elongatum *Taylor.* 20x. *D. Leave geme in* Zygodon *sp.* 20x *E. Apice of the leaf in* Neckera andina Mitt.40x *F. wavi leaf of* Neckera ehrenbergii Müll. Hal. 10x.

example,. *Zygodon quitensis* Mitt. *Zygodon fragilis* H.Rob and *Lepyrodon tomentosus* (Hook.) Mitt. which are shared in the four study areas. Other species such as *Zygodon reinwardtii* (Hornsch.) A. Braun, *Neckera andina* Mitt, *Thuidium peruvianum* Mitt. and *Sematophyllum swartzii* (Schwägr.) W.H. Welch & H.A. can be found in Kello'qocha and Manthanay.

Order/Family	Species	1	2	3	4
Orthotrichales/ Orthotrichaceae	*Zygodon quitensis* Mitt.	x	x	x	x
	Zygodon fragilis H.Rob.	x	x	x	x
	Zygodon reinwardtii (Hornsch.) A. Braun	—	x	x	—
	Zygodon sp.	—	x	x	x
	Orthotrichum elongatum Taylor	x	x	—	—
	Orthotrichum sp.	x	—	—	—
Hedwigiaceae	*Braunia cirrhifolia* (Mitt.) A. Jaeger	—	x	x	x
Leucodontales/Neckeraceae	*Neckera ehrenbergii* Müll. Hal.	x	x	—	—
	Neckera andina Mitt.	—	x	x	—
Dicranaceae	*Campylopus areodictyon* (Müll. Hal.) Mitt.	x	—	x	x
	*Chorisodontium mittenii (*Müll. Hal.) Broth.	—	—	x	—
Bryaceae	*Bryum subapiculatum* Hampe	x	—	—	—
	Bryum andicola Hook.	—	—	—	x
Bartramiaceae	*Bartramia potosica* Mont.	—	—	x	—
	Bartramia sp.	—	—	x	
Hypnales/Brachytheciaceae	*Brachythecium sp.*	x	x	—	x
	Brachythecium occidentale (Hampe) A. Jaeger	x	x		x
Thuidiaceae	*Thuidium peruvianum* Mitt.	—	x	x	—
Sematophyllaceae	*Sematophyllum swartzii* (Schwägr.) W.H. Welch & H.A. Crum	—	x	x	—
Dicranales/Pottiaceae	*Syntrichia andicola* (Mont.) Ochyra	x	—	—	x
	Leptodontium tricolor (R.S. Williams) R.H. Zander	x	x	x	x
	Syntrichia sp.	—	—	—	x
	Daltonia trachyodonta Mitt.	—	x	x	
	Didymodon challaense (Broth.) R.H. Zander.	x	—	—	—
Hookeriales/Daltoniaceae	*Daltonia* sp.	x	—	—	x
Leucodontales/ Leptodontaceae	*Morphospecies (1)*	x	x	—	—
Lepyrodontaceae	*Lepyrodon tomentosus* (Hook.) Mitt.	x	x	x	x

Table 1.
Species of mosses in the forests of Canchacancha (1), Kontorkayku (2), Kelloq'ocha (3), and Manthanay (4).

Figure 9 shows an important beta diversity for the 4 *Polylepis* forests in terms of species, genera, and families, where the Manthanay forest has the least diversity for each taxon.

In the box plot, (**Figure 10**) the richness of species of the four forests is shown graphically, we can observe that there are atypical data for the Kellococha and Manthanay forests, an important aspect to point out is that the box of this last forest presents a great difference when compared to the Canchacancha and Contorkayku

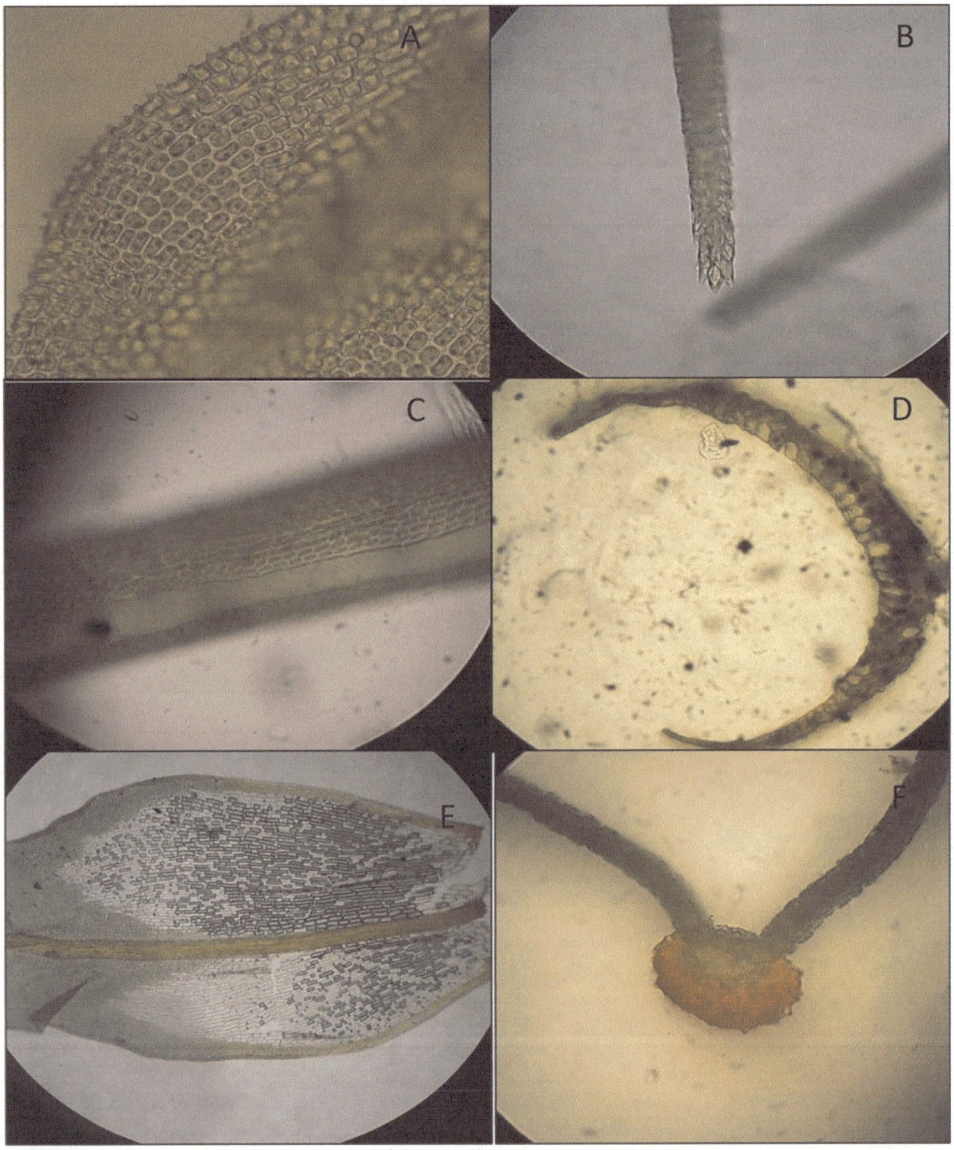

Figure 8.
A. Cells with two papillae in Thuidium peruvianun Mitt.40x *B. leave apex in* Campylopus areodictyon (Müll. Hal.) mitt. 40x; *C. involute leave in* Chorisodontium mittenii (Müll. Hal.) broth. *40x D. Clorocistos and hialocistos, phyllidia of* Chorisodontium mittenii (Müll. Hal.) broth *40x. E. Cancellinae cells in* Leptodontium tricolor (R.S. Williams) R.H. Zander 40x *F. cross-sectional of leaf in* Syntrichia andicola (Mont.) *40x.*

boxes, that is why we say that the Manthanay forest presents a difference in terms of the richness of moss species.

In the box plot, (**Figure 11**) the richness of species there is no significant difference in species richness between living and dead arboretums. The floristic composition between living and dead arboretums is different; however, it was observed that in dead arboretums the species richness is equal to that of living arboretums.

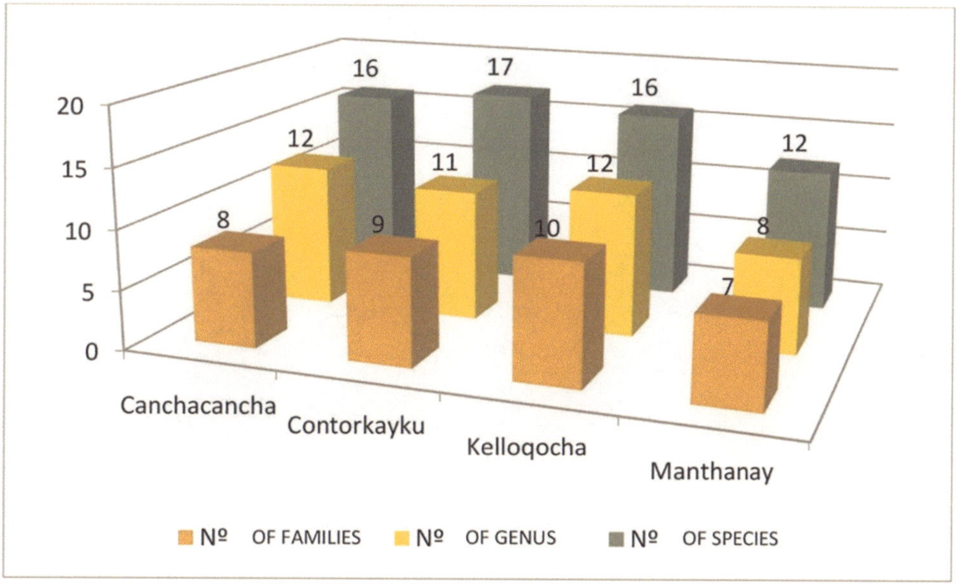

Figure 9.
Distribution of families, genera, and morphospecies in the four forests of Polylepis *spp.*

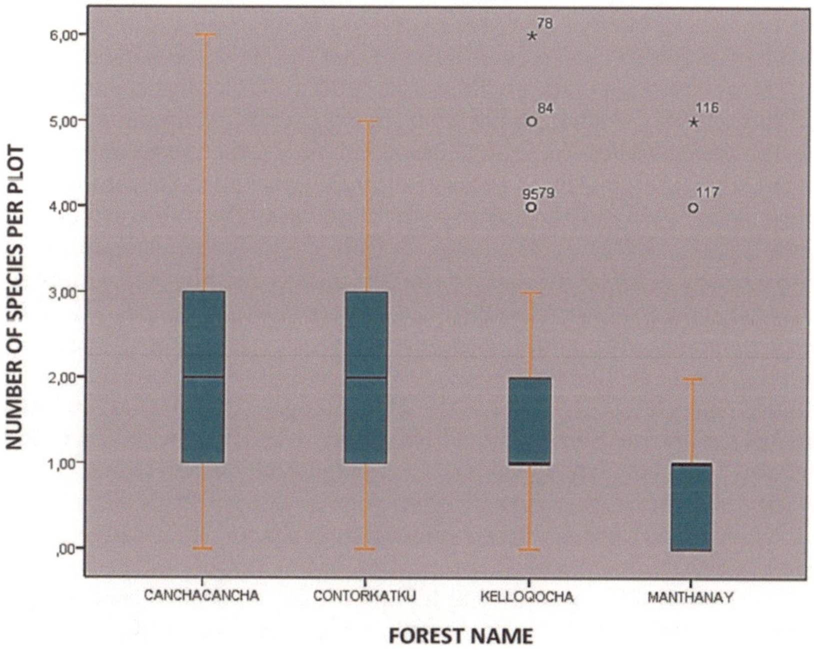

Figure 10.
Box and whisker plot for species richness per plot of each forest.

7. Discussions

Acurio [11] carried out an exhaustive study of the diversity of mosses in the Wiñay Wayna – Machupicchu area – Peru, finding a total of 129 species of mosses in 72

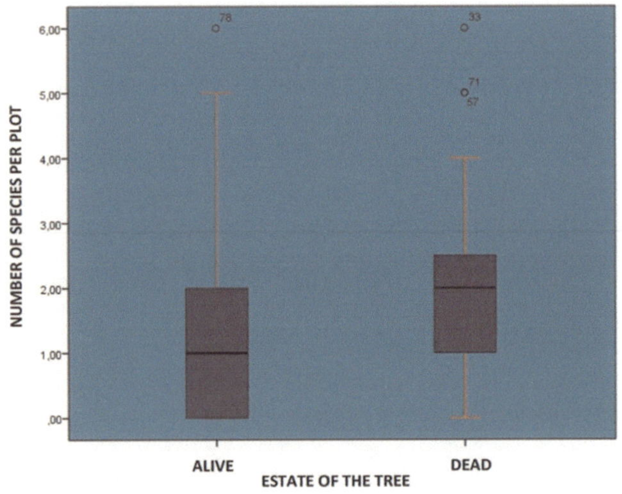

Figure 11.
Box and whisker plot for species richness per plot of each estate of the tree.

genera and 29 families. Making a comparison with the present study we have that: Acurio collected mosses in an altitude range of 2500 to 3100 m, on the other hand, this study was carried out in an altitude range of 4300 to 4800 m, and in investigations carried out by Churchill et al. [3], it was found that the elevation zone with the highest number of moss species is between 2000 and 3000 m. Lithophytic, epiphytic, and terrestrial mosses were evaluated; however, in the present study, only epiphytic mosses were taken, and according to Churchill et al. [3], the most common substrate among mosses is soil, followed by rocks and finally epiphytes. Therefore, the ecological conditions for Wiñay Wayna seem to be much more favorable, which allows the existence of a greater diversity of species. Despite the differences regarding the study area, it is necessary to highlight that, of the 129 species cited by Acurio, three of these species were also reported for this study: *Zygodon reinwardtii* (Hornsch.) A. Braum, *Thuidium peruvianum* Mitt., and *Sematophyllum swartzii* (Schwägr.) W.H. Welch & H.A. Crum.

Fuentes & Churchill [17] found, in the Madidi region (Bolivia), a total of 369 species in 168 genera and 54 families, of which seven species are shared with the present study: *Bartramia potosica* Mont., *Chorisodontium mittenii* (Müll. Hal.) Broth., *Braunia cirrhifolia* (Mitt.) A. Jaeger, *Orthotrichum elongatum* Taylor, *Z. reinwardtii* (Hornsch.) A. Braum, *Leptodontium tricolor* (R.S. Williams) R.H. Zander, *Thuidium peruvianum* Mitt. To understand the proportion of the number of species that were obtained in this study, it is necessary to highlight that the studied area comprises approximately 30,000 km^2, likewise, the climate is varied and includes mountain ranges that reach 6000 m, low mountain ranges, wide and narrow valleys, and plains above 200 m. Gradstein et al. [16] stated that in the neotropics, the diversity of mosses depends on the heterogeneity of the habitats, together with the vegetation zoning provided by the topographic relief. Likewise, Churchill et al. [18] discovered that, if the topographic relief is poor, the diversity of mosses is low. Starting from this premise, we can infer that the richness of the bryoflora in the study area is due to the geographical and ecological conditions that the study area presents (topographical relief, edaphology, climate, vegetation, etc.)

DOI: http://dx.doi.org/10.5772/intechopen.108284

For its part, Carhuapoma [13] carried out the study of mosses of the Historical Sanctuary of the Pampa de Ayacucho in an altitude range of 3350 to 4100 m.a.s.l., registering the Pottiaceae as the most representative family with 10 species, followed by Orthotrichaceae, Brachytheciaceae, and Bartramiaceae with six species, which shows that the results are similar to those reported in the present investigation.

It was believed that the adaptation of *Polylepis* did not allow the development of epiphytes on trunks (bolus, low canopy) and branches because the bark is constantly being renewed [19], but despite this difficult condition, lichens colonize and settle on branches and trunks of *Polylepis* trees, to later allow the development of mosses, liverworts and other vascular plants by ecological succession.

8. Conclusions

Studies of mosses in the high mountains of Peru, particularly in *Polylepis* forests, are still very scarce, however, we can see that beta diversity is high, with large differences in the composition of species in each forest, which is interesting and needs further studies.

Acknowledgements

We express our gratitude to Dr. Steven P. Churchill for the certification of the determined species, as well as to the Center for Research and Production of Food and Medicinal Mushrooms of the Faculty of Sciences of the National University of San Antonio Abad of Cusco for the ease of laboratory and use of stereo microscopes. We are grateful to ECOAN-Peru for helping with the logistics of the fieldwork.

Funding

Fieldwork was funded by the Leverhulme Trust, UK, the University of Zurich, the SNSF [grant P2ZHP3_161988], and the DFG [grants HE 6726 5/1; JU 2748/6–1].

Conflict of interests

The authors do not incur conflicts of interest.

Ethical/legal aspects

The authors declare that they have not incurred any violation of an ethical or legal nature.

Author details

Marisol Saji Saire[1], María E. Holgado Rojas[1*], Mitsy Diaz P.V. Sylvester[1] and Steven Paul Sylvester[2]

1 National University of San Antonio Abad of Cusco, Cusco, Peru

2 College of Biology and the Environment, Nanjing Forestry University, Nanjing, China

*Address all correspondence to: encarnacion.holgador@unsaac.edu.pe

References

[1] Fjeldsa J, Kessler M. "Conserving the Biological Diversity of *Polylepis* Woodlands of the Highland of Perú and Bolivia" a Contribution to Sustainable Natural Resource Management in the Andes. Copenhagen, Denmark: NORDECO; 1996

[2] Churchill SP, Aldana MC, Opisso J, Morales T. Familias y géneros de los Musgos de los Andes Tropicales. Editorial La Rosa. Santa Cruz de la Sierra, Bolivia Financiado por National Science Foundation Missouri Botanical Garden. 2020. p. 8

[3] Churchill SPN, Sanjines A, Aldana C. Catálogo de las Briofitas de Bolivia: La Diversidad, Distribución y Ecología. La Rosa Editorial, Santa Cruz de la Sierra, Bolivia: Museo de Historio Natural Noel Kempff Mercado y Missouri Botanical Garden; 2009

[4] Menzel M. Preliminary checklist of the mosses of Perú. Journal of the Hattori Botanical Laboratory. 1992;**71**:175-254

[5] Opisso Mejía JA. Contribución al Conocimiento de los Musgos Pleurocárpicos de la Provincia de San Ignacio. Cajamarca, Lima, Perú: Universidad Nacional Mayor de San Marcos; 2001

[6] Opisso Mejía JA. Briófitos de Pomahuaca. Cajamarca-Perú. Arnaldoa. 2003;**10**(2):2,7-14

[7] Opisso MJA, Churchill SP. Bryophytes from the environs of Yanachaga Chemillén National Park, Department of Pasco, Peru. The Bryologist. 2008;**111**(2): 310-317

[8] Galiano WH. The Flora of Yanacocha, a Tropical High Andean Forest in Southern Perú. A Thesis for the Degree of Master of Science. St. Louis, U.S.A: University of Missouri; 1990. p. 270

[9] Tupayachi HA. Flora de la Cordillera del Vilcanota. Arnaldoa. 2005;**12**(1–2): 126-144

[10] Galiano W, Tupayachi A, Nuñez M. "Flora del Valle del Cusco" *Historia Natural del Valle del Cusco*. 1 era Edición. Peru: Sociedad Protectora de la Naturaleza Cusco; 2005. pp. 197-230

[11] Acurio L. ;Musgos de la zona de Wiñay Wayna, Santuario Histórico de Machupicchu Tesis de Investigación Carrera Profesional de Biología. Peru: UNSAAC; 2002

[12] Huallparimachi QG, Molina IY, Holgado Rojas ME, Saji SM. Briofitas del Santuario Histórico de Machu Picchu. UNESCO; 2016

[13] Carhuapoma Soto JZ. Musgos en el Santuario Histórico de la Pampa de Ayacucho, distrito Quinua, provincia Huamanga. Ayacucho, Peru, 2017; 2019 Tesis para obtener el título profesional de bióloga en la especialidad de ecología y recursos naturales

[14] Holdridge LR. Life Zone Ecology. San José, Costa Rica: Tropical Science Center; 1967 [Traduccion del ingles por Humberto Jiménez Saa: Ecología Basada en Zonas de Vida, la. Ed. San José, Costa Rica: IICA; 1982]

[15] Gradstein SR, Nadkarni NM, Kromer T, Holz IN, Noske N. A protocol for rapid and representative sampling of vascular and non-vascular epiphyte diversity of tropical rain forests. Selbyana. 2003;**24**:105-111

[16] Gradstein SR, Churchill S, Salazar Allen N. A guide to the bryophytes of

tropical America. Memoirs of the New York Bolivia Garden. 2001;**86**:1-577

[17] Fuentes A, Churchill S. Catálogo de las briófitas de la región de Madidi, Bolivia. Ecología en Bolivia. 3ra Edición. 2005. pp. 170-198

[18] Churchill SP, Griffin D III, Lewis M. Diversidad de Musgos en los Andes Tropicales. New York, USA: The New York Botanical Garden; 1995. pp. 335-346

[19] Purcell J, Brelsford A, Kessleer M. "The World's highest Forest" a better understanding of the properties of Andean queuña woodlands has major implications for their conservation. American Scientist. 2004;**92**:454-461

Chapter 3

Bryophyta around Syamsudin Noor International Airport, South Kalimantan, Indonesia

Yuninda Riyana, Sasi Gendro Sari and Gunawan Gunawan

Abstract

Research on biodiversity of bryophyta around Syamsudin Noor International airport has never been reported. The study purposed the types of mosses around the Syamsudin Noor International Airport. A purposive sampling method was used to collect bryophyta and then environmental parameters were directly measured in the field. Moss identification referred to the book *A Handbook of Malesian Mosses* volume 1 *Sphagnales to Dicranales* and volume 2 *Leucobryaceae to Buxbaumiaceae*. Morphological characteristics of mosses were observed at leaf parts, stem, and rhizoids. The results showed 7 species of mosses consisting of *Barbula javanica, Hyophila apiculata, Philonotis hastata, Ectropothecium falciforme, Fissidens atroviridis, Calymperes tenerum*, and *Hyophila involuta*. The dominant species was *Barbula javanica*. The range of environmental parameter valued air temperature 28.1–34.1°C, light intensity 20–35 lux, humidity 74–84%, and soil pH 6.5–7. The results of this study are preliminary data that can be used for further research and determine the changing conditions of the environment around Syamsudin Noor International Airport.

Keywords: airport, bryophyta, diversity, bioindicator, environmental

1. Introduction

Moss comes from two words, namely Bryon which means moss and Python which means moist or wet [1]. Several species of mosses which are still classified as thallus form faults and have never found further differences, but mosses are found which and body composition is close to the characteristics of higher plants, namely having stems, leaves, and rhizoids (still in a simple form) [2]. The characteristics of mosses have parts consisting of capsules, seta, leaves, stems (transition from thallus to cormus), and rhizoids (mosses do not yet have true roots) [1]. Moss is recorded to have approximately 18,000 species in the world and is a low-level plant group that ranks second after higher plants. While in Indonesia recorded 1500 species of moss that have been found. This is related to mosses that can live in humid areas because Indonesia is a tropical region which has humid areas so that it supports a lot of moss to live. Moss grows in clumps that cover the surface of the substrate such as weathered wood, litter, tree trunks, rocks, and leaves. Moss has several roles and benefits, one of which is an important component that plays a role in air balance and nutrient cycles in an area, especially tropical rain forests. Lichens can act as pioneer plant nutrients in rich locations. Moss can also act as a bioindicator of environmental pollution in an area or area contaminated with pollutants [3]. This is because mosses have the opportunity to change the environment [4].

This is one of the interesting reasons in this study, the area that became the focus around the airport area which was originally a peat swamp and no human activity with developments that changed environmental conditions into a crowded location. In addition, the conditions around the airport have a view of humid conditions. Environmental changes are also affected by the biodiversity in the area. Thus, this study aims to reveal the diversity of mosses around the Syamsudin Noor international airport area, Banjarbaru, South Kalimantan, Indonesia.

2. Bryophyta around Syamsudin Noor airport

This research was carried out in 5 stages, the first activity was collecting moss samples using the purposive sampling method by taking the complete part of the moss and storing it to maintain moisture and safety. Second, measurement of environmental parameters includes air temperature, humidity, light intensity, and soil pH. Third, identification is carried out with reference books, namely A Handbook of Malesian Mosses Volume 1 Sphagnales to Dicranales [5] and A Handbook of Malesian Mosses Volume 2 Leucobryaceae to Buxbaumiaceae [5]. Fourth, the morphological characters of mosses consisting of leaves, stems, and rhizoids were observed using a stereo microscope. Fifth, data analysis was carried out descriptively. Several types of moss found around the Syamsudin Noor airport area are:

2.1 *Barbula javanica* Dozy and Molkenboer

B. javanica is found on rock and soil substrates. The most common moss found in various open or closed places in the research area. *B. javanica* can also be found in moist or wet soil [6]. Live in groups and side by side with other mosses when found in the field. Individual length at the time of measurement reached 4 mm. Rhizoid when observed shaped like a thread and firmly attached to the substrate. The stem when observed is almost invisible because it is covered by an arrangement of leaves. The stem shape is erect, brownish, unbranched, and can reach 7 mm in height [7]. The leaves when observed are green, lanceolate in shape, flat edges, and tapered ends, the leaf length reaches 0.6 mm, and shrivels when dry. The leaves curl when dry, are green, lanceolate, blunt tip, flat edge [6], and mother leaf bones are brownish in color [7]. The shape of leaf cells when observed is rectangular and the size of leaf cells reaches 2 m. The cells in the middle of the strands that make up the midrib are rectangular and the cells at the edges of the strands are rectangular [6]. In the field observations found gametophyte and sporophyte phases. The sporophyte phase was found to be seta with a length of 1.8 mm and a capsule with a length of 0.4 mm, calyptra with a length of 0.5 mm, the shape of the calyptra was long and tapered, the spores were round (**Figure 1**).

The environmental parameters of *B. javanica* consisted of air temperature of 28.1°C, light intensity of 35 lux, humidity of 84%, and pH 7. The results of environmental parameters are still in the range of good moss growth. Distribution of *B. javanica* is found in Tropical Asia, Central Africa, Cape Verde Islands, Guinea, Nigeria, Arabian Peninsula, Oman, Tropical Central America, United States, Florida, Hawaii, and Australia [8].

Figure 1.
Barbula javanica: *A. Soil substrate, B. Colonies, C. Individual mosses, D. Leaf anatomy, E. Spores.*

2.2 *Hyophila apiculata* Fleisher

Mosses are found in the soil substrate and grow in groups. *H. apiculata* can also be found on rock substrates [5]. At the time found living side by side with other mosses. The genus Hyophila is found in an open environment [4]. *H. apiculata* is on average small in stature [5] and looks like scales when viewed from above [3]. Individual length when measured reaches 5.5 mm. Morphological observations in the laboratory, rhizoids resemble threads and stick firmly to the substrate. Brownish rhizoids measuring less than 1 mm [4]. The stem when observed is covered by leaves so that it is not visible. Measuring less than 3 mm [5]. In the Hyophila genus, the upper stem is green and the lower part is red to reddish brown or dark green, reaching 1 cm high and sometimes branching [4]. The leaves are less than 2 mm in size and are ovate to spatulate [5]. When found with light green leaves, the shape of the leaves looks lanceolate or can be spathulate, flat edges, and pointed ends. The leaves can be lanceolate and light green in color [9] and the leaf edges are flat, the tips are tapered and curved, and the leaf arrangement is alternate (**Figure 2**) [10].

In the genus Hyophila, the leaf bones are large and erect from base to tip [4]. The shape of the leaf cells when observed was square and the cell size was 11 m. The sporophyte phase was not found in the field. Seta can grow to a length of 8 mm and there is a capsule and calyptra growing at the tip of the seta [9]. *H. apiculata* has environmental factors when found including air temperature 28.1°C, humidity of 84%, light intensity 35 lux, and pH 7. These results are in accordance with a good growth range in mosses. The distribution of *H. apiculata* is distributed in Malesia [5].

Figure 2.
Hyophila apiculata: *A. Soil substrate, B. Colonies, C. Individual mosses, D. Leaf anatomy.*

2.3 *Philonotis hastata* Wijk and Margadant

P. hastata is found in soil substrates. In addition to soil, it can also be found on rock substrates [3] and tree substrates [10]. When found living in groups and side by side with other mosses. This moss looks like a sponge and is thick [10]. Observations in the laboratory individual length reaches 4 mm. Rhizoid when observed as threads firmly attached to the substrate. It is smooth and small [3], and in the genus Philonotis rhizoid it is thin [11]. The stem when observed is almost invisible because it is covered with leaves. The shape of the stem is upright and 2–8 mm long [3]. The leaves are green to yellowish according to [11] *P. hastata* is bright green to yellowish green. Spiral leaf arrangement and curls when dry [3]. Leaf length when measured reached 1.1 mm, lanceolate shape, flat edge, and pointed tip. The shape is lanceolate, the tip is pointed, the leaf edge is flat [3]. The shape of the leaf cells when observed was rectangular and the length of the leaf cells was 11 m. The sporophyte phase was not found in the field. The environmental parameters of *P. hastata* consisted of air temperature of 28.1°C, humidity 84%, light intensity 35 lux, pH 7. These results are in accordance with a good growth range in mosses. The distribution of *P. hastata* is found in Mexico, Central and South America, Africa, Asia, Australia, Brazil, and Oceania (**Figure 3**) [12].

2.4 *Ectropothecium falciforme* Dozy and Molk

E. falciforme was found in the base of the tree substrate. Other substrates commonly grown by *E. falciforme* are rock, tree roots, and leaf epiphytes and the habitat for growing is usually in a slightly open environment [4]. Lives creeping on the substrate and side by side with other mosses when found in the field. Individual length when measured reaches 9 mm. Rhizoid when observed like a thread firmly attached to the substrate. The observed stems spread on the substrate according to research sources that the stems spread on the substrate and are covered with leaves [3]. The length of

Figure 3.
Philonotis hastata: *A. Soil substrate, B. Colonies, C. Individual mosses, D. Leaf anatomy.*

the stem reaches 7.5 cm, the branches are slightly pinnate and narrow at the ends, are regular, and look dense because the arrangement of the leaves overlaps each other [4]. The leaves when measured have a length of up to 0.6 mm. The length of the leaves can reach 1 mm [4]. The leaves are light green, lanceolate in shape, flat edges, and pointed ends. The leaves are yellowish green, shiny, the leaves are round lanceolate, the edges of the leaf tips are jagged [4]. The leaves are small and have a pointed tip [3]. Leaf cell shape when observed like smooth muscle and leaf cell length reaches 3 m (**Figure 4**).

Figure 4.
Ectropothecium falciforme: *A. Tree substrate, B. Colonies, C. Individual mosses, D. Leaf anatomy, E. Spores.*

Gametophyte and sporophyte phases are found in this moss. The sporophyte structure consists of seta at the time of measurement which is 11 mm long and brown in color. The seta are brown and have a capsule at the end of the seta [3]. The capsules when measured were 0.9 mm long and brown in color. Calyptra on *E. falciforme* observations were not seen. The shape of the spores when observed is spherical and 1.4 m in diameter. The environmental parameters of *E. falciforme* consisted of air temperature of 34.1°C, humidity 79%, light intensity 20 lux, and pH 7. The results from both locations were still in a good growth range for mosses. Distribution in Indonesia is in Sumatra, Java, Lombok, Kalimantan, and Sulawesi [4].

2.5 *Fissidens atroviridis* Besch

F. atroviridis was found on soil substrates. *F. atroviridis* can grow on moist and shady soil and rocks [5]. In addition, it can be found on weathered tree substrates [13]. Coexists with other mosses when found in the field. Small in stature [5] and looks like a comb from above [3]. Individual length when measured reaches 7 mm. The individual length of *F. atroviridis* can reach 10–15 mm [14]. The structure when dry will bend [5]. Rhizoid when observed like a thread firmly attached to the substrate. The rhizoid structure in the Fissidens genus is smooth or has palillae [11]. The stem when observed is almost invisible because it is covered by leaves. The leaves that have been observed are dark green. *F. atroviridis* is dark green in addition, there is also a pale yellowish to transparent dull green color (**Figure 5**) [5].

The leaves when measured were 1.9 mm long. The leaf length can reach 2.5 mm [5]. The arrangement of the leaves is pinnate and very tight so it looks piled up [9]. Thick leaf margins [14]. The shape of the lancet, flat edge, and pointed tip are in accordance with the characteristics of the research found [9]. The shape of the leaf cells when observed was hexagonal and the length of the leaf cells was 1.1 m. The hexagonal leaf cells are irregular, very thin-walled, and the cells are large [14]. In *F. atroviridis*

Figure 5.
Fissidens atroviridis: *A. Soil substrate, B. Colonies, C. Individual mosses, D. Leaf anatomy, E. Spores.*

gametophyte and sporophyte phases were found. The sporophyte phase in this moss consists of seta which is brownish green when observed and the measured length reaches 3.2 mm and that the length of the seta can reach 4 mm [5]. The capsule when observed was green and the length of the capsule when measured was 0.4 mm. The capsules are symmetrical [14]. Caliptra was measured to reach 0.5 mm and was brown in color, the shape of the spores when observed was round, and the diameter of the spores when measured was 0.8 m. The environmental parameters of *F. atroviridis* consisted of air temperature of 32.2°C, humidity 84%, light intensity 22 lux and pH 6.5. The results from both locations were still in a good moss growth range. Distribution of *F. atroviridis* is found throughout Indo-Malaya from India to New Guinea [5].

2.6 *Calymperes tenerum* Müll. Hall

C. tenerum was found on the base of the tree substrate. In addition to tree substrates, *C. tenerum* can be found in rock, soil, and rotting wood [15]. Coexists with other mosses when found in the field. Individual length when measured reaches 4.6 mm. The length of *C. tenerum* rarely grows more than 7 mm and is very small [5]. Rhizoid when observed like a thread firmly attached to the substrate. The stem when observed is almost invisible because it is covered with leaves. The stem grows upright and acrocarp [15], and has a length of up to 1 cm [14]. The leaves when measured have a length of 1.4 mm, green color, lanceolate shape, flat edge, and the tip has a capsule and immediately grows together with the seta. The length of the leaves is between 1.5–2 mm, dull green in color, does not or almost does not widen below the base of the branch, the leaves are almost uniform, and the leaf edges are intact [5]. The shape of the leaf cells when observed was round and the length of the leaf cells reached 1 m. *C. tenerum* leaf cells have 1 layer, the shape is not always rounded and has thin walls in the upper branches [14]. *C. tenerum* found gametophyte and sporophyte phases. The sporophyte phase observed consisted of seta with a length of up to 1.9 mm and green in color. The capsules when observed were cylindrical and brown in color according to [14] which stated that the capsules were cylindrical. The capsules look like thorny spheres, this is in accordance with research [15] which mentions capsules like thorns that characterize this moss. The capsule when observed measuring 0.1 mm. The spores are spherical in shape and 0.9 m in diameter. The environmental parameters of *C. tenerum* consisted of air temperature of air temperature 34.1°C, humidity 79%, light intensity 20 lux, and pH 7. The results of environmental parameters from both locations were still in a good growth range for mosses. The distribution of *C. tenerum* is found around the coast of tropical Asia, India, the Pacific islands to Hawaii, the lowlands of Malesia, especially near the sea (**Figure 6**) [14].

2.7 *Hyophila involuta* Jaeger

H. involuta is found on rock substrates and grows in groups. Other substrates can be found in soil, logs, and walls. Small upright stature with a height of 1 mm, but can grow up to 1.5 cm [5]. At the time found living side by side with other mosses. Living in an open environment [4]. The rhizoids when found are thread-like and firmly attached to the substrate. Rhizoid is thin [11] and looks wrinkled [5]. The stems when observed are covered by leaves so that they are not visible. In the genus Hyophila, the upper stem is green and the lower part is red to reddish brown or dark green, reaching 1 cm high and sometimes branching [4]. The leaves when observed are green. This is in accordance with [16] that the leaves are green but can be colored to dark yellowish green. Leaves spread when moist and curl when dry [5]. The genus Hyophila has alternate leaf

Figure 6.
Calymperes tenerum: *A. Tree substrate, B. Colonies, C. Individual mosses, D. Leaf anatomy, E. Spores.*

sitting [4]. Observations in the laboratory leaf length reached 0.6 mm, lanceolate shape, flat edge, and pointed tip. This is in accordance with [4] that the genus Hyophila has flat leaf edges, grows upright from the base to the tip of the leaf, and according to [16] pointed leaf tips. The mother leaf bone is strong, brownish or reddish, protruding, and grows to the tip of the leaf [14]. The shape of leaf cells when observed is square according to [5] that the upper lamina cells are rectangular, the cell walls are solid, and the cell width reaches 7–9 m. Leaf cell length when measured reached 1.1 m. At the base of the leaf is colorless, small, and the cells are rectangular and thin-walled (**Figure** 7) [5].

H. involuta did not find a sporophyte phase, while according to library sources the size of the seta reached 7 mm to 1.5 cm [11], red to brown at the base, pale at the top, cylindrical capsule shape [5]. Environmental parameters of *H. involuta* from air temperature have a value of 28.1°C, humidity 84%, light intensity 35 lux, and pH 7 which is still in good growth for mosses. Distribution is found in tropical Asia as generally found throughout the Malesia region and the genus Hyophila is the only genus scattered in this region and spread in sub-tropics such as Polynesia, northern Australia [5]. Distributions elsewhere include Africa, Asia (China, India, Himalayas, Japan, Malaysia, Philippines, Sri Lanka, Taiwan, Thailand), Europe, Oceania, North and South America [14].

2.8 Environmental parameters

Environmental conditions had a significant influence on mosses. The differences of each species of mosses to environmental factors will affect the level of adaptation, species composition, and distribution of mosses [14]. Mosses that live in the tropics can live at temperatures above 25°C [14] with air humidity above 50% [17]. While good light intensity for moss growth can reach 1000 lux [4] and for good soil pH for moss growth it ranges from 4.9 to 8.3 [18]. Based on this, the environment around the Syamsudin Noor airport area is still in the range of environmental conditions that support the growth of moss considering that the place is an open and urban area where there are many activities that cause changes in environmental conditions, one

Figure 7.
Hyophila involuta: *A. Stone substrate, B. Colonies, C. Individual mosses, D. Leaf anatomy.*

of which is pollution. While in open areas direct sunlight penetrates into the ground, this affects the humidity of the area. In addition, the open area has homogeneous vegetation and only has a few trees as windbreaks. It is different when in forest areas which have heterogeneous vegetation with a wide and closed canopy so as to minimize sunlight entering the lower part of the forest [19]. Environmental changes can occur at any time, so this initial data is very important to know environmental changes in the future, given that mosses can be used as an indicator of environmental change.

3. Conclusion

Bryophyta in the vicinity of Syamsudin Noor International Airport, South Kalimantan, Indonesia, found 7 species of moss which are found both in open and closed places and the most dominant species is Barbula javanica. Considering that the airport is an open area, the dominant moss growing in this area has a good standard of living, as can be seen from environmental parameters including air temperature, humidity, light intensity, and soil pH. In addition, this study is preliminary data that can be used to determine other types of moss around Syamsudin Noor International Airport.

Acknowledgements

The author thanks the reviewers who have devoted their time and expertise to review to ensure the quality of this paper. Besides that, thank you to the institution where the research has facilitated the completion of this research and thanks to the funders who have help the writer in the form of material so that this research is going well until it's finished.

Additional information

Parts of this chapter were first published in the original Indonesian article by the same authors: Riyana Y, Gendro Sari S, Gunawan G. Bryophyta di Sekitar Kawasan Bandar Udara Internasional Syamsudin Noor Kalimantan Selatan [Internet]. Vol. 2, Jurnal Jejaring Matematika dan Sains. Universitas Palangka Raya; 2020. p. 36–40. Available from: http://dx.doi.org/10.36873/jjms.2020.v2.i2.402.

Author details

Yuninda Riyana*, Sasi Gendro Sari and Gunawan Gunawan
Departement of Biology, Lambung Mangkurat University, Banjarbaru, South Kalimantan, Indonesia

*Address all correspondence to: yunindariyana82@gmail.com

References

[1] Lukitasari M. Mengenal tumbuhan lumut (*Bryophyta*) deskripsi, klasifikasi, potensi dan cara mempelajarinya. Magetan: CV. Ae Media Grafika; 2018

[2] Tjitrosoepomo G. Taksonomi tumbuhan (*Scyzophyta, Thallophyta, Bryophyta, Pteridophyta*). Yogyakarta: UGM Press; 2011

[3] Raihan C, Nurasiah, Zahara N. Keanekaragaman Tumbuhan Lumut (*Bryophyta*) di Air Terjun Peucari Bueng Jantho Kabupaten Aceh Besar. Prosiding Seminar Nasional Biotik. 2018;**2018**:440-451

[4] Endang T, Jumiati, Pramesthi IAD. Inventarisasi Jenis-Jenis Lumut (Bryophyta) di Daerah Aliran Sungai Kabura-Burana Kecamatan Batauga Kabupaten Buton Selatan. Jurnal Biologi Tropis. Mei 2020;**20**(2):161-172

[5] Eddy A. A Handbook of Malesian Mosses. 1st & 2nd ed. London: British Museum (Natural History); 1988

[6] Salamah Z, Sasongko H, Zulianti E. Diversity of *Bryophyte* in Selarong Cave Area, Bantul, Yogyakarta. Indonesian Journal of Biology and Education. 2019;**2**(1):35-39

[7] Köckinger H, Kučěra J, Hofmann H, Müller N, Amann G. Barbul consanguinea discovered in Switzerland and Austria, with a revision of former European records of *B. Indica*. Herzogia. 2010;**25**(1):61-70

[8] Sollman P. Studies on *Barbula consanguinea* (Thw. & Mitt.) Jaeg. sensu Eddy, a pan-tropical species. Tropical Bryology. 2000;**19**:17-23

[9] Febrianti GN. Identifikasi Tumbuhan Lumut (Bryophyta) di Lingkungan Universitas Jember serta Pemanfaatanya sebagai Buku Nonteks [skripsi]. Jember: Universitas Jember; 2015

[10] Zahara M. Jenis-jenis tumbuhan lumut (Bryophyta) di stasiun penelitian Soraya kawasan ekosistem Leuser sebagai referensi mata kuliah botani tumbuhan rendah [skripsi]. Banda Aceh: Universitas Islam Negeri Ar-Raniry; 2009

[11] Rahmawati YR. Penyusunan buku eskripsi ragam lumut di lingukungan SMAN 1 Pundong sebagai media pembelajaran biologi. Jurnal Prodi Pendidikan Biologi. 2018;**7**(1):35-43

[12] Oliveira-da-Silva FR, Ilkiu-Borges AL. Flora das cangas da Serra dos Carajes, Para, Brasil: Bartramiaceae. Rodriguesia. 2016;**6**(5):1125-1128

[13] Zahara M. Jenis-jenis tumbuhan lumut (Bryophyta) di stasiun penelitian Soraya kawasan ekosistem Leuser sebagai referensi mata kuliah botani tumbuhan rendah [skripsi]. Banda Aceh: Universitas Islam Negeri Ar-Raniry; 2019

[14] Sumadjiko H, Vitara PE. Diversity of Bryophyte at junior and senior high schooles environment in Yogyakarta city for supporting the implementation of curriculum 2013 in applicative biology learning. AIP Conference Proceedings. 2022:1-11

[15] Fajriah R. Keanekaragaman lumut (Bryophytes) pada berbagai substrat di kawasan Sungai Pucok Krueng Raba kecamatan Lhoknga kabupaten Aceh Besar sebagai referensi praktikum ekologi tumbuhan [skripsi]. Banda Aceh: Universitas Islam Negeri Ar-Raniry; 2018

[16] Kavitha R, Sathish SS, Vijayakanth P, Palani R. Thamizharasi T, Vimala A.

Diversity and distribution of the family Pottiaceae (Bryopsida) from Pachamalai Hills, Eastern Ghats of Tamil Nadu. IJIRT. 2018;**4**(8):84-90

[17] Adhitya F, Ariyanti NS, Djuita NR. Keanekaragaman Lumut Epifit pada Gymnospermae di Kebun Raya Bogor. Floribunda. 2014;**4**:212-217

[18] Purbasari YA, Akhmadi AN. Keanekaragaman Bryophyta di Dusun Sumbercandik Kabupaten Jember. BIOMA: Jurnal Biologi dan Pembelajaran Biologi. Mar 2019;**4**(1):90-100

[19] Khotimperwati L, Rahadian R, Baskoro K. Perbandingan Komposisi Tumbuhan Lumut Epifit pada Hutan Alam, Kebun Kopi dan Kebun Teh di Sepanjang Gradien Ketinggian Gunung Ungaran, Jawa Tengah. Bioma. 2015;**17**(2):83-93

Section 3

Interactions: Bryophytes and Other Organisms

Chapter 4

Bryophilous Agaricomycetes (Fungi, Basidiomycota): A Review to Brazil

Cassiane Furlan-Lopes, Alice Lemos Costa,
Marines de Avila Heberle, Fernando Augusto Bertazzo-Silva,
Ana Luiza Klotz-Neves, Kamille Rodrigues Ferraz,
Guilherme Henrique Mueller, Marina de Souza Falcão,
Jorge Renato Pinheiro Velloso and Jair Putzke

Abstract

Bryophilous fungi have at least one stage of its life cycle linked to Bryophytes. There are few studies in relation to their taxonomy and ecology all around the world, including Brazil. The Agaricomycetes (Basidiomycota) have gained prominence worldwide and contained several species of economic interest. Based on a bibliographic review and discussion about identification methods and experimental models on this association a species list of bryophilous/Agaricomycetes found in Brazil was elaborated. In the works found among the techniques used to identify effective fungi/Bryophytes associations it can be cited: phylogenetics analysis, optical and electron microscopy, and cultivation experiments. In Brazil, four orders of Agaricomycetes (Basidiomycota), belonging to Agaricales, Boletales, Hymenochaetales, and Polyporales, with 33 species were found associated to Bryophytes in the literature. Information of the worldwide distribution of Brazilian muscicolous species and application of these groups were realized associating edibility, toxicity, and others. It was noted that in this country there is a scarcity of scientific knowledge of this subject, that needs to be better understood in terms of ecology and taxonomy.

Keywords: bryophytes, mosses, liverworts, Agaricales, Hymenochaetales, Polyporales

1. Introduction

Bryophylous or muscicolous fungi can be defined as those that have at least one stage of its life cycle linked to Bryophytes (mosses, liverworts and hornworts). Bryophytes do not produce nutrient-rich storage structures or specialized transport tissues rich in photosynthetic products as vascular plants, but fungal pathogenesis of mosses is being reported with increasing frequency [1]. The relationship between these two groups is indicated as important to the land colonization by plants [2–4]. Studies of bryophyte/fungal symbioses have also provided powerful insights into the origin and evolution of mycorrhizal associations in land plants [4]. Palaeontologic researchers found fungal structures of Glomeromycota associated with young bryophytes in Ordovician sediments 460 and 400 million years old [5, 6].

Mycorrhizal fungi that inhabit symbiotically healthy tissues of terrestrial plants using organs of absorption [7] inhabiting the rhizoids and/or thalli of liverworts and hornworts were reported associated to different fungi phylla, as Ascomycota, Glomeromycota, and Basidiomycota [4, 8, 9]. Ascomycota has been better studied in this field of knowledge than Basidiomycota [8–11]. Among the Basidiomycota, Agaricomycetes present a total of 19 known orders: Agaricales, Amylocorticiales, Atheliales, Auriculariales, Boletales, Cantharellales, Corticiales, Geastrales, Gloeophyllales, Gomphales, Hymenochaetales, Hysterangiales, Jaapiales, Phallales, Polyporales, Russulales, Sebacinales, Thelephorales, and Trechisporales [12]. With a worldwide distribution, the Agaricomycetes have gained prominence since the class contain several species of economic interest [13–15].

Many of the Agaricomycetes orders are mentioned in different studies about Bryophilous fungi, such as Agaricales, Hymenochaetales, and Polyporales [16–18]. Different structures were developed by fungi to parasite the Bryophytes [1]. Morphological, molecular, and in vitro experiments have shown that mosses such as *Sphagnum*, *Polytrichum* and *Hylochomium* are often associated with Agaricomycetes [16, 17, 19]. A biotrophic trophic mode, in which the fungi species can degrade plant cell walls and lignin, cleaving sucrose to glucose, was recently suggested to occur on 15 species of Hymenochaetales [18].

In Brazil, the are few studies citing the association between Agaricomycetes and Bryophytes [16, 20]. This "gap" in these subjects to science often occurs due to a lack of cooperation between Bryologists and Mycologists [1]. This almost absence of scientific knowledge on this subject is considered as a barrier to better understanding this relation in terms of ecology and taxonomy.

This study as a revision of the knowledge generated up to date to Brazil in this area is an outline of the main employed methods used to identify the interactions of bryophilic mushrooms, as well as a revision of the data of occurrence in the world, and perform a list of Brazilian bryophilous Agaricomycetes. The data presented here are proposed as a starting point to call for more mycologists and bryologists to join the efforts to better understand the Fungi-Bryophyte relationship.

2. Materials and methods

The main employed methods used to identify the interactions of bryophilic mushrooms were illustrated, also, studies in the world with bryophilous Agaricomycetes among 1980 to 2022 were revised. A list of bryophilous Agaricomycetes found in Brazil, with their distribution was made, using bibliographic research available in: Google Scholar, Scopus and Scielo. The nomenclature follows according to the Index Fungorum [12]. The distribution also was obtained from Global Biodiversity Information Facility (GBIF) platform, with the filter "Preserved Specimen" and "Reflora" (Flora e Funga do Brasil). The states of Brazil were named with their respective acronyms: Acre - AC, Alagoas - AL, Amapá - AP, Amazonas - AM, Bahia - BA, Ceará - CE, Distrito Federal - DF, Espírito Santo - ES, Goiás - GO, Maranhão - MA, Mato Grosso - MT, Mato Grosso do Sul - MS, Minas Gerais - MG, Pará - PA, Paraíba - PB,

Paraná - PR, Pernambuco - PE, Piauí - PI, Roraima - RR, Rondônia - RO, Rio de Janeiro - RJ, Rio Grande do Norte - RN, Rio Grande do Sul - RS, Santa Catarina - SC, São Paulo - SP, Sergipe - SE, Tocantins - TO.

3. Methods used to identify the interactions of bryophilic mushrooms

The most used methods for the identification of bryophyta/Agaricomycetes associations are: optical and electronic microscopy, molecular and phylogenetic analyses, and *in vitro* culture experiments (**Figure 1**). Initially, it is necessary to identify the site of mushroom/bryophyte association, such as non-photosynthetic regions like rhizoids, or photosynthetic regions like the thalli or leaf structures. As for example, an optical microscopy analysis was used to identify the fungi *Chromocyphella muscicola* (Fr.) Donk in association with bryophytes, reporting this species usually known from the Northern Hemisphere in Brazil for the first time [21].

With the preparation of slides with KOH (5%), it is possible to visualize the structures of the mushrooms, especially the hyphae which are sometimes linked to the bryophyte cells. In the scanning electron microscope, slides were prepared as usually with reagents that can also be used to identify the association between Bryophytes and Agaricomycetes fungi [1]. As a differential, when analyzing the species *Sphagnum fuscum* Klinggräff with this methodology, it was possible to visualize the rudimentary appressoria that mechanically facilitated the entry into the cells of photosynthetic structures, belonging to the bryophilic species *Glomus mosseae* (T.H. Nicolson & Gerd.) Gerd. & Trappe [1]. An illustrative schematic of the step-by-step of these techniques is shown in **Figure 1A**.

The phylogenetic analyses can be made to detect the feeding and ecological habits using gene portions (ITS 1–2 and 5.8S rRNA) DNA extractions, sequencing and with subsequent bioinformatic analysis [19]. Analysis performed with *Mycena* sp. and *Galerina* sp. showed close evolutionary relationships with *Dicranum* sp. and *Hylocomium* sp. [19]. Key findings include that *Galerina* sp. showed a preference to associate with senescent, rather than photosynthetic tissues, and thus ancestral saprotrophic habit. On the other hand, *Mycena* sp. showed colonization in both tissues, and therefore ancestral parasitic habit [19]. In general, phylogenetics is performed in several steps: (I) Material preparation; (II) DNA extraction; (III) Sequencing; and (IV) Sequence analysis by bioinformatics [22]. This results in four advantages: an independent framework for clade construction; a well-supported statistical basis, as the sites of an alignment integrate matrices of different sizes; a low incidence of putative homeoplasies compared to morphological characters; and the implementation of evolutionary models applied independently to each base [23]. An illustrative schematization of these main steps is shown in **Figure 1B**.

In vitro culture experiments are performed to analyze the ecology of interactions and resistance of bryophytes. The bryophytic fungi are part of a diffuse group, often only detected by molecular analyses [24]. The basidioma emerges at specific periods, temperature and humidity, which can make it difficult to visualize between the photosynthesizing or non-photosynthesizing structures of bryophytes [25]. The mycelial phase is the most predominant fungal phase, and this structure can be visualized under microscope when associated to bryophytes. When growing the species *Atrichum androgynum* (Müll. Hal.) A. Jaeger in culture medium, an association with the fungi *Arthrobotrys oligospora* Fresen., was visualized which is known to capture nematodes [24]. This was only possible due to the growth of the fungi in

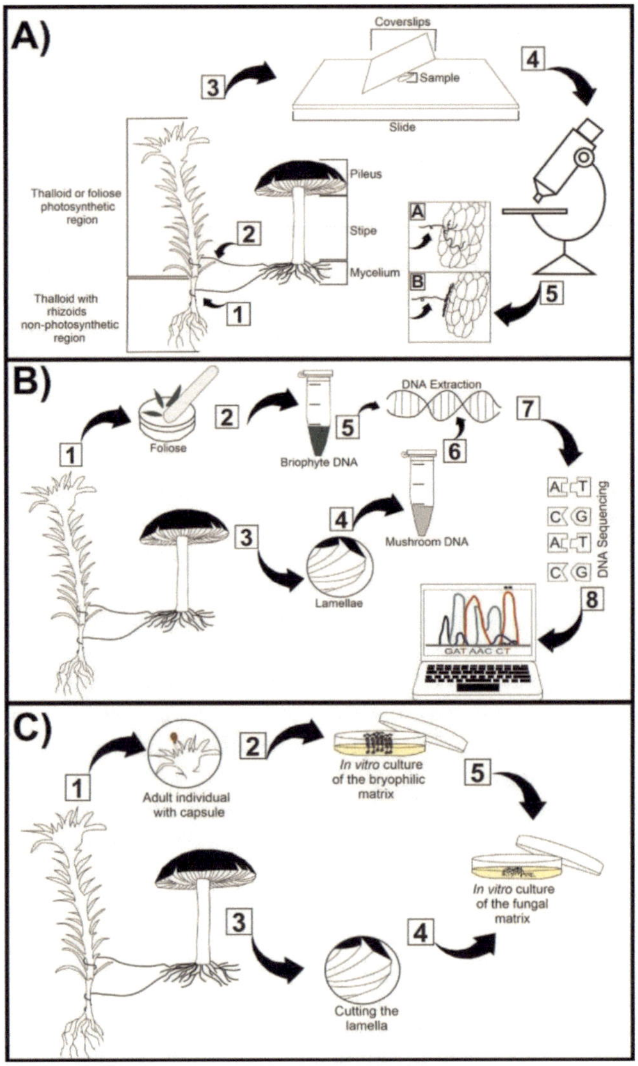

Figure 1.
Methodology frequently used for detection of Fungi-bryophyte associations (FBA). (A) Optical microscopy: association site, non-photosynthetic (1) or photosynthetic (2); slide cuts and use of reagents with cover by coverslip (3); microscope observation (4); FBA endophilic (5-A) or exophytic (5-B) structures. (B) Steps of molecular analyses of FBA: preparation of material for DNA extraction in bryophytes (1 and 2); preparation of material for DNA extraction in mushrooms (3 and 4); use of reagents for DNA extraction (5 and 6); sequencing of gene portions of interest (7); analysis of the sequences by bioinformatics techniques (8). (C) Visualization of FBA in culture medium: a mature bryophyte containing the capsule is isolated (1), and disposed in culture medium (2); the basidiomata (3), when lamellar region is cut and the structure is placed in culture medium (4); when the fungi is not visible, the hyphae grow in the culture medium starting from the bryophytes and can be isolated and cultured separately for species identification (5).

culture medium, since it was not visible among the collected bryophytes [24]. Sometimes the structures of the fungi can be detected so that with the aid of microscopy, tweezers, and accessories the fungi can be isolated and grown separately in usual culture medium (like PDA). An illustration of the methodological steps mentioned above is shown in **Figure 1C**.

4. Studies about bryophilic mushrooms

Bryophilic mushrooms have been known for a long time, their habitats are well known (swamps, moss-covered tree trunks and mounds). In the 1980's the works about bryophilous fungi around the world include optical microscopy observations, in both Ascomycota and Basidiomycota. In 1981, the fungal species *Lyophyllum palustre* (Peck) Singer was found in mosses and it was constantly associated with necrotic areas of the species, and relationships with bryophytes discussed. Apparently less aggressive species of associated mushrooms were obtained from pure cultures made of *Sphagnum capillaceurn* (Weiss) Schrank, isolating *Lyophyllum palustre* and *Galerina paludosa* (Fr.) Kühner [26]. Research was carried out in 1987 on bryophilic fungi found in samples from the main herbaria in central and northern Europe, identifying ascomycetes associated with the bryophyte *Polytrichum sexangulare*: *Bryochiton heliotropicus* Döbb., *Bryochiton perpusillus* Döbb., *Lizonia sexangularis* Döbb. & Poelt., *Protothelenella polytrichi* Döbb. & Mayrh., *Gloeopeziza interlamellaris* Döbb. and *Hymenoscyphus norvegularis* Döbb [27].

Bryophytes are involved in a variety of competitive, parasitic, symbiotic, mutualistic and not yet specified interactions with fungi [28]. Some of Bryophilous species have very specific substrates such as *Galerina paludosa* found only in *Sphagnum* swamps, *Cyphellostereum laeve* was found on polytrichoid mosses in coastal regions, in addition to *Rickenella fibula* (Bull.) Raithelh. and *Rickenella setipes* which descriptions reported that a mound formed by *Alnus glutinosa* roots, covered by bryophytes served as substrate for the development of the species [29, 30]. In the 1990s, the main genera of bryophytic fungi were well known: *Galerina*, *Omphalina*, *Rickenella*, *Hypholoma*, *Mycenella* and *Psilocybe*, whose collection area was more widespread, knowing that their nature is more delicate they must be collected with a considerable amount of their substrate of origin [30]. Some islands were targets of ecological studies, such as South Georgia (Southern South America) and Iceland where the bryophyte substrate used by some Agaricomycetes was reported. Collections focused further south of the island of South Georgia were found to have species of the genus *Galerina*, *Gerronema*, *Phaeogalera* and *Hypholoma* [31]. Experiments about growth symbiosis using *Laccaria* spp. on different substrates, including *Sphagnum*, showed that when *Sphagnum* was mixed with vermiculite it was beneficial for the development of the *Laccaria* species and its symbiotic effects [32]. Another research reports that *Sphagnum* and other bryophytes can increase the presence of macrofungi in the substrate [33]. Studies of capture and evasion of nitrogen in soils with suspension of mosses verified that rhizomes of bryophytes have covering links with hyphae of Basidiomycetes fungi, forming a sheath around the rhizoids. This is important since the soil alone cannot handle nitrogen excess, nor accumulating without moss [34].

Research related to Agaricomycetes fungi and mosses among the years 2000 to 2010 cover mainly themes involved in the taxonomy, phylogeny, distribution, diversity, and classification of these organisms. Based on morphological characteristics and phylogenetic analysis, it was proposed that the *Omphalina giovanellae* Bres. systematic position is better maintained in another genus, making the mushroom that grows between mosses and low grass belonging to the genus *Clitopilus* (Fr. ex Rabenh.) P. Kumm [35]. *Multiclavula ichthyiformis* Nelsen, Lücking, L. Umaña, Trest & Will-Wolf, was identified as a new basidiolichen from Costa Rica with terricolous habits that grows with bryophytes [36]. *Psathyrella laurentiana* A.H. Sm and *Omphalina philonotis* (Lasch) Quél. (currently classified as *Arrhenia philonotis* (Lasch) Redhead,

Lutzoni, Moncalvo & Vilgalys) had their occurrence linked to bryophytes and/or peatlands, suggesting a more specialized niche for some basidiomycetes [37]. The occurrence of decomposing species of the genus *Galerina* Earle and *Coprinus* Pers. in bryophyte swamps including *Polytrichum alpestre* Hoppe in South Georgia Island [38], previously reported [32] can suggest a niche specialization in some species of these genera. *Chromocyphella muscicola* (Fr.) Donk was reported for the first time in Turkey [39], a species that grows near or on mosses, or even parasitizes them while alive [40]. The occurrence in association with mosses is used as identifying character for *Maireina callostoma* (Pilát) W.B. Cooke when using an identification key of this genus [41], showing that associations between Bryophytes and Fungi is an important character also to the taxonomy of Fungi.

Phylogenetic analysis helped to understand the bryophilic habits and high concentration of basidiolichens in Hygrophoraceae (Basidiomycota, Agaricales), suggesting a predisposition of these fungi to change their mutualist nutrition associated with photobionts to saprotrophic [42]. The occurrence of *Marasmius epidryas* Kühner, currently classified as *Rhizomarasmius epidryas* (Kühner ex A. Ronikier) A. Ronikier & Ronikier, among mosses was recorded in cold areas of Canada, Denmark and Russia highlights the occurrence of this relationships in cold and harshest areas [43].

Among 2015–2022 most analyses focused on more complex analyses about bryophilic mushrooms. Association among *Sphagnum* mosses and 26 species of Agarics and Boleti (Agaricomycetes, Basidiomycota) were found in Ukraine, whereas *Galerina cerina*, *G. paludosa*, *G. sphagnicola*, *Hypholoma elongatum*, *H. udum,* and *Tephrocybe palustris* could be considered as closely associated by substrate links [44]. In 2018, it was demonstrated that many bryophilous Hymenochaetales have values of stable isotope indicating ectomycorrhizal habits or a biotrophic cluster indicative of parasitism or an endophytic lifestyle [18].

The mycorrhizal-like associations, diversity and distribution of fungal associations in bryophytes, as between liverworts of the Jungermaniidae and Marchantiidae with Basidiomycota clades were explored by bibliographic review [4]. Although several works have been published reporting the occurrence of Agaricomycetes fungi growing among mosses, the importance of more specialized studies on the subject is highlighted. Part of these works did not perform microscopic and phylogenetic analysis in order to prove the relationship between Fungi and Bryophytes, but only reported the growth of certain species among mosses. The use of microscopic and phylogenetic analysis will contribute not only to prove the association, but also to understand the phylogeny and evolution of these organisms.

5. Brazilian Agarycomycetes growing with Bryophytes

In Brazil, a total of 33 species from four orders of Agaricomycetes were reported growing with mosses and/or liverworts. Agaricales was the most representative order, with 10 families and 14 genera divided into 20 species. Hymenogastraceae shows a greater number of species associated with *Sphagnum*, although many of the other species of mosses associated with Agaricomycetes have not been identified in the literature. Hymenochaetales presented three families and seven genera, divided into nine species. This order stands out by the diversity mosses and liverworts associated with fungi, including *Rickenella fibula*, which has a feeding habit specialized in bryophytes [18]. Polyporales presented three families and three genera, with several species of mosses and liverworts. Boletales presented only one species growing next to mosses.

List of Agaricomycetes fungi reported to Brazil:

Agaricales Underw.

Agaricales incertae sedis

Rimbachia arachnoidea (Peck) Redhead, Can. J. Bot. 62(5): 878, 1984. ≡ *Mniopetalum bisporum* Singer, Darwiniana, 14: 10, 1966.

Ecology and importance: Growing gregarious on unidentified mosses [45, 46]. *R. arachnoidea* could also be confused with *Rimbachia bryophila* (Pers.) Redhead. Both species are parasitic on mosses and differ in not having venose hymenophore [47].

Distribution: In Brazil it is found in RS (**Figure 2**) [46]. It is found also in Spain, Germany, Switzerland, Norway, Finland, Sweden, Austria, Estonia, Netherlands, Denmark, Belgium, Canada, Argentina, New Zealand, United States of America, Turkey, Poland, among others [47–49].

Chromocyphellaceae Knudsen.

Chromocyphella muscicola (Fr.) Donk, *Persoonia* 1(1): 95, 1959.

≡ *Arrhenia muscicola* (Fr.) Quél., *Fl. mycol. France* (Paris) 33, 1888.

Ecology and importance: Grow among unidentified mosses and in lichens [21, 50].

Distribution: In Brazil it is found in MG (**Figure 2**) [21, 50]. It is also found in Switzerland, Germany, Spain, Portugal, Australia, Austria, Norway, New Zealand, Netherlands, Cuba, Turkey, among others [39, 48].

Clavariaceae Chevall.

Clavaria fragilis Holmsk., Beata Ruris Otia Fungis Danicis, 1: 7, 1790.

Ecology and importance: Growing in the ground with unidentified mosses [51]. This species is edible [14], showing antioxidant activity [52].

Distribution: In Brazil it is found in RS, SC, and PR (**Figure 2**) [51]. It is also found in Russian Federation, United States of America, Norway, Finland, Sweden, Italy, Switzerland, Spain, Japan, Germany, Canada, Netherlands, Puerto Rico, Australia, South Africa, among other [48, 53–55].

Hymenogastraceae Vittad.

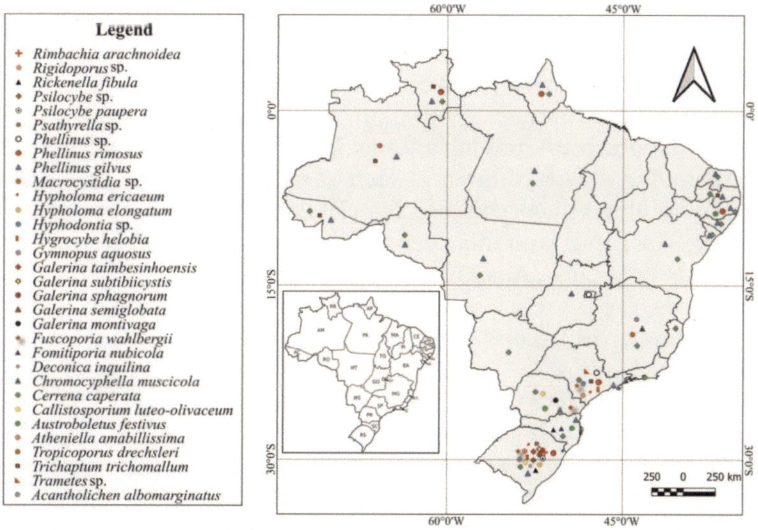

Figure 2.
Distribution of bryophilous species in Brazilian states/regions.

Galerina montivaga Singer, *Nova Hedwigia,* 29: 306, 1969.

Ecology and importance: Growing gregarious in unidentified moss fields and on humus [56–58].

Distribution: In Brazil it is found in PR (**Figure 2**) [56–58]. It is found also in the United States of America, Slovakia and Argentina [48].

Galerina semiglobata Singer, *Lilloa*, 26: 147, ('1953'), 1954.

Ecology and importance: Forming dense groups on *Sphagnum* that, in some points, are burned. In Brazil it was found in RS (**Figure 2**) [20, 58].

Distribution: This species is endemic to Brazil and is found in Rio Grande do Sul state [20, 58].

Galerina sphagnorum (Pers.) Kühner, *Encyclop. Mycol.*, 7: 179, 1935. Sanctioned by Fries.

Ecology and importance: Grow gregarious in *Sphagnum* [20, 58]. In Romania, this species is in the red list in the category "near threatened" [59]. In Poland, *G. sphagnorum* was found associated with eight different species of mosses, *Polytrichum commune, S. centrale, S. fallax, S. cuspidatum, S. flexuosum, Sphagnum magellanicum, S. palustre and S. papillosum* [60].

Distribution: In Brazil it is found in RS (**Figure 2**) [20, 58]. It is also found in Russian Federation, Finland, United States of America, Estonia, Sweden, Austria, Switzerland, Japan, Belgium, Spain, Canada, among others [48, 59].

Galerina subtibiicystis Singer, *Lilloa*, 26: 146 ('1953'), 1954.

Ecology and importance: They grow scarcely among the peat bogs of *Sphagnum* moss [20, 58].

Distribution: This species is endemic to Brazil and is found in RS (**Figure 2**) [20, 58].

Galerina taimbesinhoensis Singer, *Lilloa*, 26: 148 ('1953'), 1954.

Ecology and importance: Growing exclusively on *Sphagnum* moss [20, 58].

Distribution: This species is found in RS (**Figure 2**) [20, 58]. *G. uchumachiensis* Singer is considered a synonym of *G. taimbesinhoensis* [61]. It is also found in the Hawaiian Islands [62].

Psilocybe paupera Singer, Sydowia, 9 (1–6): 404, 1955.

Ecology and importance: Growing gregarious, attached to the stalks of the moss *Sphagnum* [58, 63, 64]. Belongs to the Red List of Macrofungi of China [65].

Distribution: In Brazil, it is found in RS (**Figure 2**) [58, 63, 64]. It is found also in Germany, Costa Rica and China [48, 65].

Psilocybe sp. (Fr.) P. Kummer.

Ecology and importance: Growing among *Sphagnum* in open marshes [20]. This genus can be found in diverse substrates such as soil, dung, wood, and mosses [66].

Distribution: In Brazil the bryophyte associated specimen was found in RS (**Figure 2**) [20]. The genus contains over 150 species distributed worldwide [67]. Occurs worldwide in Mexico, Australia, Canada, Sweden, Germany, United Kingdom, Spain, Netherlands, Costa Rica, Iceland, France, Argentina, New Zealand, Colombia, Russian Federation, Japan, among other countries [48].

Hygrophoraceae Lotsy.

Hygrocybe helobia (Arnolds) Bon, Docums Mycol. 6(no. 24): 43, 1976. ≡ *Hygrocybe miniata* (Fr.) P. Kumm., *Der Führer in die Pilzkunde*: 112, 1871.

Ecology and importance: It was found growing on soil, often between unidentified mosses, and are generally gregarious [25, 68].

Distribution: In Brazil it is found in RS and SP (**Figure 2**) [25, 68]. It was also recorded in Finland, Russian Federation, Switzerland, Sweden, Spain, Austria, Germany, Colombia, Costa Rica, Canada, among other countries [48].

Acantholichen albomarginatus Dal-Forno, Marcelli & Lücking, Mycologia 108(1): 43, 2016.

Ecology and importance: found on the edge of Nebular forest, by the road, on dense vegetation on road side banks, growing on unidentified mosses and liverworts [69].

Distribution: Endemic to Brazil, found in MG (**Figure 2**) [69].

Psathyrellaceae Vilgalys, Moncalvo & Redhead.

Psathyrella sp.

Ecology and importance: Growing in woods at mountains and among unidentified mosses [20]. *Psathyrella* has about 400–600 species, and molecular studies suggest its separation into several others [58]. This genus presents species that have antibacterial diterpenoids [70].

Distribution: In Brazil the bryophyte associated specimen was found in RS (**Figure 2**) [20]. In the world it is reported to the United States of America, Norway, Finland, Congo, Australia, Germany, Russian Federation, Estonia, Spain, New Zealand, Austria, Sweden, Denmark, Poland, Japan [48].

Strophariaceae Singer & Smith.

Hypholoma elongatum (Pers.) Ricken, Die Blätterpilze 1: 250, 1915. ≡ *Psilocybe uda* (Pers. ex Fr.) Gillet, Hyménomycètes (Alençon): 586, 1878.

Ecology and importance: it was found growing attached to the stalk of *Sphagnum* moss, away from trees [20].

Distribution: In Brazil it is found in the RS (**Figure 2**) [20]. It is a cosmopolitan species and is reported to the United States of America, Sweden, Norway, Australia, Russian Federation, Mexico, Japan, Switzerland, Spain, Colombia, New Zealand, Germany, South Africa, Poland, Costa Rica, Argentina, France, Iceland, Bolivia, Indonesia, China, Ukraine, Cameroon, Portugal, Republic of Korea, Peru, Paraguay, among other countries [48].

Hypholoma ericaeum (Pers.: Fr.) Kühner, Bull. Trimest. Soc. mycol. Fr., 52: 23, 1936.

Ecology and importance: Growing in wet and sandy soils among unidentified mosses and grasses [71, 72]. Found in peat bogs among or near peat mosses, but also in wet meadows [73].

Distribution: In Brazil it is found in RS and SP (**Figure 2**) [71, 72]. It is also reported to Spain, United States of America, Poland, Belgium, Norway, Netherlands, Denmark, Russian Federation, Australia, Austria, France, and Greenland [48].

Deconica inquilina (Fr.) Pat. ex Romagn., Revue Mycol., Paris 2(6): 244, 1937. ≡ *Psilocybe muscorum* (P.D. Orton) M.M. Moser, in Gams, Kl. Krypt.-Fl., *Ed.* 3 (Stuttgart) 2b/2: 239, 1967.

Ecology and importance: Growing among an unidentified moss in sandy soil [71].

Distribution: In Brazil it is found in RS (**Figure 2**) [71]. It is also reported to Poland, Estonia, Spain, Iceland, United States of America, Austria, Colombia, Italy, Mexico, Portugal, Russian Federation, Denmark, Norway, Sweden, Finland, Canada, Switzerland, Belgium, Germany, among other countries [48].

Biannulariaceae Jülich.

Callistosporium luteo-olivaceum (Berk. & M.A. Curtis) Singer, *Lloydia* 89: 117, 1946. ≡ *Callistosporium luteofuscum* Singer, *Lilloa*, 26: 115 ('1953'), 1954.

Ecology and importance: It grow on decaying wood and between the moss *Sphagnum* [20, 46, 57]. It was also reported with saprotrophic habits, on angiosperm wood and growing naturally on rich, deep and moist soils [74].

Distribution: In Brazil it is found in RS and PR (**Figure 2**) [20, 46, 57]. It is reported also to the United States of America, Japan, Canada, Switzerland, Spain, Austria,

Costa Rica, Netherlands, Sweden, Australia, Norway, Bolivia, China, Colombia, Czechia, among other countries [48, 72].

Omphalotaceae Bresinsky.

Gymnopus aquosus (Bull.) Antonín & Noordel., in Antonín, Halling & Noordeloos, Mycotaxon 63: 363, 1997 ≡ *Collybia dryophila* (Bull. ex Fr.) Kummer var. *oedipus* Quél., *Fl. mycol. France (Paris)*: 226, 1888. Bas.: *Agaricus dryophilus* Bull. ex Fr., *Herb. Fr. (Paris)*, 10: 434, 1790. ≡ *Marasmius dryophilus* (Bull. ex Fr.) Karsten, *Finl. Nat. Folk*, 48: 103, 1889.

Ecology and importance: It grow in a humid open environment away from trees, associated with *Sphagnum* [20, 46]. It presents β-glucan with antioxidant activity [75].

Distribution: In Brazil it is found in RS (**Figure 2**) [20, 46]. It is also reported to Denmark, France, Germany, Sweden, Finland, Spain, Austria, Russian Federation, Switzerland, Norway, United Kingdon of Great Britain, Netherlands, Estonia, among other countries [48, 75, 76].

Mycenaceae Overeem.

Atheniella amabillissima (Peck) Redhead, Moncalvo, Vilgalys, Desjardin & B.A. Perry, Index Fungorum 14: 1, 2012. ≡ *Mycena amabilissima* (Peck) Sacc., *Syll. Fungorum*, 9: 37. 1891. Bas.: *Agaricus amabilissimus* Peck, *Rep. (Annual) Trustees State Mus. Nat. Hist.*, New York, 39: 39 ('1886'), 1887. ≡ *Prunulus amabilissimus* Murrill, *North Am. Flora*, 9: 324, 1916.

Ecology and importance: Growing among unidentified mosses [46, 77].

Distribution: In Brazil it is found in RS (**Figure 2**). In the world it is reported to the United States of America, Canada, Finland and Argentina [46, 48, 77, 78].

Macrocystidiaceae Kühner.

Macrocystidia sp.

Ecology and importance: Associated with an unidentified moss species [79].

Distribution: In Brazil the bryophyte associated specimen was found in AM (**Figure 2**) [79]. The genus is reported also to Sweden, Denmark, Norway, Spain, Germany, Switzerland, New Zealand, Finland, Austria, United States of America, Japan, Belgium, Canada, Poland, Estonia, Iceland, Mexico, France, Italy, Netherlands, Australia, Republic Democratic of Congo, Czechia, United Kingdom of Great Britain and Northern Ireland, among other countries [48].

Boletales.

Boletaceae Chevall.

Austroboletus festivus (Singer) Wolfe, *Biblthca Mycol.*, 69: 92, 1980 ('1979'). ≡ *Porphyrellus festivus* Singer, *VI. Lilloa*, 26:57–159, 1953.

Ecology and importance: Growing among unidentified mosses in Restinga Forest [80–82].

Distribution: In Brazil it is found in PE, PR, and SC (**Figure 2**) [80–82]. It is also reported to Guyana [48, 83].

Hymenochaetales.

Hymenochaetales *incertae sedis.*

Trichaptum trichomallum (Berk. & Mont.) Murrill, *Bull. Torrey bot. Club* 31(11): 608, 1904.

Ecology and importance: Growing associated with the moss *Entodon beyrichii* (Schwaegr.) C. Muell. Int the Cerrado biome [16]. It is an edible species [13].

Distribution: In Brazil it is found in SP, RO, PB, AC, and AM (**Figure 2**) [16, 48]. It is reported also to Mexico, Peru, Costa Rica and United States of America [48].

Hyphodontiaceae.

Hyphodontia sp.

Ecology and importance: Growing in Atlantic Forest associated with the mosses - *Fabronia ciliaris* (Brid.) Brid. *var. polycarpa* (Hook.) Buck, *Isopterygium tenerum* (Sw.) Mitt., *Sematophyllum subpinnatum* (Brid.) Britt., *Syrrhopodon africanus* (Mitt.) Par. *subsp. graminicola* (Williams) Reese [16]. It has been found also associated with liverworts - *Chonecolea doellingeri* (Nees) Grolle, *Harpalejeunea molleri* (Steph.) Grolle, *Lejeunea flava* (Sw.) Nees, *Metzgeria cf. dichotoma* (Sw.) Nees [16].

Distribution: In Brazil the bryophyte associated specimen was found in SP [16]. This genus occurs also in PA, RO, MG, RJ, PR, RS and SC (**Figure 2**) [84]. In the world, is reported to Sweden, Spain, Denmark, Norway, Estonia, Switzerland, Germany, Australia, United States of America, New Zealand, Finland, Poland, Austria, Ukraine, Belgium, Russian Federation, France, Canada, Portugal, India, Costa Rica, Italy, Romania, Islamic Republic of Iran, Réunion, Turkey, Ethiopia, United Republic of Tanzania, Japan, Argentina, Colombia, French Guiana, among other countries [48].

Hymenochaetaceae Donk.

Fomitiporia nubicola Alves-Silva, Bittencourt & Drechsler-Santos, *Mycological Progress*, 19(8): 769–790, 2020.

Ecology and importance: Growing on the living tree of *Drimys angustifolia*, among unidentified mosses [85].

Distribution: Described from Brazil, found in SC [85].

Fuscoporia wahlbergii (Fr.) T. Wagner & M. Fisch., Mycol. Res. 105(7): 780, 2001. ≡ *Phellinus wahlbergii* (Fr.) D.A. Reid, Contr. Bolus Herb. 7: 97, 1975.

Ecology and importance: Growing between mosses - *Octoblepharum pulvinatum* (Dozy & Molk.) Mitt., *Syrrhopodon prolifer* Schwaegr. var. *acanthoneuros* (C. Muell.) C. Muell., *Trichosteleum papillosum* (Hornsch.) Jaeg. [16]. It can be fuond growing also with liverworts: *Calypogeia peruviana* Nees & Mont., *Cephalozia crassifolia* (Lindenb. & Gott.) Fulf., *Cyclolejeunea luteola* (Spruce) Grolle, *Kurzia capillaris* (Sw.) Grolle, *Monodactylopsis minima* (Schust.) Schust., *Riccardia chamaedryfolia* (With.) Grolle, *Telaranea nematodes* (Gott. ex Aust.) Howe, *Zoopsis antillana* Steph. [16]. Antioxidant activity [86].

Distribution: In Brazil it is found in BA, RJ, SP, PR, RS, and SC (**Figure 2**) [16, 87]. It is also reported to New Zealand, Japan, Democratic Republic of Congo, Spain, United States of America [48].

Phellinus rimosus (Berk.) Pilát, *Annls mycol.* 38(1): 80, 1940. ≡ *Fulvifomes rimosus* (Berk.) Fiasson & Niemelä, Karstenia 24(1): 26, 1984.

Ecology and importance: Growing between mosses - *Erythrodontium squarrosum* (C. Muell.) Par., *Racopilum tomentosum* (Hedw.) Brid., *Trichostomum weisioides* C. Muell., *Campylopus cryptopodioides* Broth., *I. tenerum* (Sw.) Mitt., *Syrrhopodon gaudichaudii* Mont., *Thamniopsis incurva* (Hornsch.) Buck., *R. tomentosum* (Hedw.) Brid. [16]. Also, can be growing between liverworts - *Anoplolejeunea conferta* (Meissn.) Evans, *Aphanolejeunea* sp., *Bazzania heterostipa* (Steph.) Fulf., *Cephaloziella stellulifera* (Tayl.) Schiffn., *Drepanolejeunea mosenii* (Steph.) Bischl. *L. flava* (Sw.) Nees, *Plagiochila bunburii* Taylat. [16]. Presents cytotoxic, antitumor and antimalarial activity [15, 88].

Distribution: In Brazil it is found in "Cerrado" vegetation and secondary Atlantic Forest in SP, PE, RS, AP, and RR (**Figure 2**) [16, 89–91]. It is also reported to Australia, United States of America, Mexico, Japan, Senegal, Democratic Republic of Congo, Tanzania, Zimbabwe, Canada, Rwanda, Bahamas, Costa Rica, France, Kenya, Bangladesh, Belize, Ecuador, Spain, Gambia, India, among other countries [48].

Phellinus gilvus (Schwein.) Pat., Essai Tax. Hyménomyc. (Lons-le-Saunier): 82, 1900.

Ecology and importance: Growing between mosses - *Pyrrhobryum spiniforme* (Hedw.) Mitt., *I. tenerum* (Sw.) Mitt [16]. It can be also found growing with liverworts - *T. nematodes* (Gott. ex Aust.) Howe, *L. flava* (Sw.) Nees [16]. This is a medicinal mushroom showing antitumor activities, anti-oxidative, anti-fungal, healing, to treat stomach ache and various inflammations [92, 93].

Distribution: In Brazil it is found in SP, AC, AM, BA, GO, MT, PA, PB, PR, PE, RN, RS, RO, RR, SC, SP, and SE (**Figure 2**) [16, 94]. It is also reported to the United States of America, Mexico, Australia, Japan, Costa Rica, Democratic Republic of Congo, New Zealand, Jamaica, French Guiana, Peru, Zimbabwe, Burundi, Puerto Rico, Uganda, Argentina, Canada, Rwanda, Ecuador, among other countries [48].

Phellinus sp.

Ecology and importance - Growing between mosses - *I. tenerum* (Sw.) Mitt., *P. spiniforme* (Hedw.) Mitt., *Thamniopsis incurva* (Hornsch) Buck, *Sematophyllum galipense* (C. Muell.) Mitt, *S. subpinnatum* (Brid.) Britt., *Campylopus cryptopodioides* Broth., *C. cryptopodioides* Broth., *F. ciliaris* (Brid.) Brid. var. *polycarpa* (Hook.) Buck, *Thamniopsis incurva* (Hornsch.) Buckand [16]. It can be also found in liverworts - *Cheilolejeunea trifaria* (Reinw. et al.) Mizut., *Frullania ericoides* (Nees) Mont., *Lophocolea bidentata* (L.) Dum., *L. martiana* Nees, *Radula angulata*, Steph. *Riccardia chamaedryfolia* (With.) Grolle, *T. nematodes* (Gott. ex Aust.) Howe Steph., *D. mosenii* (Steph.) Bischl., *Microlejeunea globosa* (Spruce) Steph., *Aphanolejeunea subdiaphana* (Jovet Ast) Pócs var. *cristulata* (Schust.) Pócs, *C. doellingeri* (Nees) Grolle, *L. flava* (Sw.) Nees, *Lejeunea ulicina* subsp. *bullata* (Taylor) Schust., *L. glaucescens* Gott., *L. martiana* Nees, *L. muricata* (Lehm.) Nees [16].

Distribution: In Brazil, the bryophyte associated specimens are found in DF and SP [16], but this genus also occurs in AC, AL, AM, BA, CE, ES, MA, MT, MS, PA, PR, PB, PE, PI, RJ, RN, RS, RO, RR, SC, SP, SE (**Figure 2**) [16, 95]. The genus *Phellinus* is reported to the United States of America, Brazil, Sweden, Norway, Finland, Estonia, Australia, Russian Federation, Canada, Mexico, Costa Rica, Germany, Switzerland, Spain, Austria, Japan, New Zealand, Czech Republic, China, Argentina, Democratic Republic of Congo, Denmark, India, among other countries [48].

Tropicoporus drechsleri Salvador-Montoya & Popoff, in Salvador-Montoya, Costa-Rezende, Ferreira-Lopes, Borba-Silva & Popoff, Phytotaxa 338(1): 80. 2018.

Ecology and importance: Frequently among unidentified mosses growing on it (Pagin-Claudio et al., 2022).

Distribution: In Brazil it is found in MG (**Figure 2**) [96]. It is also reported to Argentina [48].

Rickenellaceae Vizzini

Rickenella fibula (Bull.) Raithelh., *Metrodiana* 4: 67, 1973.

Ecology and importance: Growing solitary to gregarious, inhabiting moss beds in high altitude areas (700 and 1500 m above sea level), found inhabiting humid moss beds of *Polytrichium* Hedw. and *Schizymenium* Harv. [97]. Presence of psilocybin [98].

Distribution: In Brazil, it is found in RS and SC (**Figure 2**) [46, 97]. It is also reported to United States of America, Norway, Denmark, Switzerland, Germany, Sweden, Finland, Poland, Spain, Australia, Russian Federation, Canada, New Zealand, Japan, Austria, Belgium, Iceland, Netherlands, China, Democratic Republic of Congo, Italy, among other countries [48].

Polyporales Gäum 1926

Polyporaceae Fr. ex Corda 1839

Trametes sp.

Ecology and importance: Growing between mosses - *Donnellia commutata* (C. Muell.) Buck, also can be found grown with the liverworts *D. mosenii* (Steph.) Bischl. [16].

Distribution: In Brazil, the bryophyte associated specimen is found in SP [16], but this genus also occurs in AC, AM, AP, PA, RO, RR, TO, AL, BA, PA, PB, SE, MS, MT, MG, RJ, SP, PR, RS and SC (**Figure 2**) [99]. The genus *Trametes* is reported also to the United States of America, Mexico, Australia, Norway, Japan, Sweden, Germany, Switzerland, Spain, Costa Rica, Canada, Russian Federation, Estonia, Finland, Denmark, Austria, Argentina, Jamaica, among others countries [48].

Cerrenaceae Miettinen, Justo & Hibbett 2017

Cerrena caperata (Berk.) Zmitr., Mycena 1(1): 91, 2001. ≡ *Datronia caperata* (Berk.) Ryvarden, Mycotaxon 23: 172, 1985.

Ecology and importance: Growing between mosses - *I. tenerum* (Sw.) Mitt., *Thamniopsis langsdorffii* (Hook.) Buck, also can be found grown with the liverworts - *D. mosenii* (Steph.) Bischl., *Lejeunea glaucescens* Gott. and *L. martiana* Nees. [16]. Present cytotoxic and immunomodulatory activity [87].

Distribution: In Brazil, it is found in AC, AL, AP, BA, ES, MT, MG, PA, PB, PR, PE, RJ, RN, RS, RO, RR, MS, SC, SP and SE (**Figure 2**) [16, 100]. Around the world it is found in Brazil, Costa Rica, Mexico, Panama, Democratic Republic of Congo, Cuba, Guyana, Venezuela, Trinidad and Tobago, Puerto Rico, Kenya, Cameroon, Colombia, French Guiana, Peru, Bolivia, Belize, Guatemala, Nicaragua, Argentina, United States of America, Ghana, Suriname, Tanzania, Sri Lanka, among other countries [48].

Meripilaceae Jülich 1982

Rigidoporus sp.

Ecology and importance: Growing between mosses - *I. tenerum* (Sw.) Mitt., *Thamniopsis langsdorffii* (Hook.) Buck, also can be found grown with the liverworts - *Lejeunea caespitosa* Lindenb., *L. martiana* Nees, *T. nematodes* (Gott. ex Aust.) Howe [16].

Distribution: In Brazil, the bryophyte associated specimen is found in SP [16], this genus also occurs in AC, AM, AP, PA, RO, RR, AL, BA, CE, MA, PB, PE, SE, MT, PR, RS and SC (**Figure 2**) [101]. The genus *Rigidoporus* is reported to Costa Rica, Mexico, United States of America, Australia, Germany, Estonia, Denmark, Sweden, Puerto Rico, Switzerland, Norway, Japan, New Zealand, Democratic Republic of Congo, French Guiana, Panama, Canada, among other countries [48].

6. Discussion

The bryophilous Agaricomycetes have been analyzed by optical and electron microscopy, *in vitro* cultures, DNA sequencing and phylogenetic analysis in the articles published up to now. Some works highlighted other relevant characteristics, such as edibility, toxicity, and antioxidant properties. Thirty-three species from four orders of Agaricomycetes were reported growing with mosses and liverworts in Brazil. In general, the diversity of possible associations between bryophytes and Agaricomycetes of woody basidioma stands out in small species of lamellate fungi. Moreover, most studies including Brazilian bryophilic species do not approach this interaction satisfactorily, and usually the associated bryophytes are not identified, or when they are identified, it is not described how the association is really occurring.

Bryophyte-Fungi associations sometimes present mutually beneficial symbiosis with bidirectional exchange of resources between partners, i.e., to be mycorrhizal-like even in the absence of true roots in bryophytes [4]. Also, it has not been investigated whether opportunistic parasitism of damaged or stressed bryophytes occurs [1]. In Brazil, about 1524 species of bryophytes were recorded, distributed in 117 families, divided into 11 species of hornworts, 633 liverworts, and 880 mosses [102]. These plants display a higher diversity and a greater number of species in areas of higher elevations and with less anthropic activity [102]. However, bryophytes may grow on different substrates, such as corrugated iron roofs, invertebrates, among others [16]. Among the types of substrates colonized by bryophytes there is a predominance of corticolous, followed by terricolous, rupicolous, and epixilous [102], without the mention of Fungicolous Bryophyte lifestyle.

In our study, the majority of mosses/liverworts and fungi growing together shows occurrences from high and damp places, in the Atlantic Forest, Amazonia, Caatinga, and Pampa biomes. Some species found in our revision, such as, *Rickenella fibula* and *Gerronema sphagnorum*, among others, occurs only on mosses [18, 20, 58]. Many of the identified bryophytes occur on identified fungi [16], such as of *Phellinus*, *Fuscoporia,* and *Hyphodontia.* Also, the mosses usually grew abundantly on the dead trunks on which the fungi were found [16]. Although bryophytes can colonize different environments, studies reveal that the majority of bryophyte species must have a preference for a single type of substrate [103]. The appearance of bryophytes must be influenced by several abiotic factors, as light availability, moisture, and water [102]. The fungi also need moist environments to complete their life cycle but light is not needed by all species [25].

The abiotic conditions are usually found in mountainous tropical areas, as humid climate with rains distributed throughout the year, ample variation of temperatures from the lowland to the high mountains, high rainfall, and topography, all factor enabling greater number of microhabitats [102]. Despite bryophytes can produce defense mechanisms, in response to fungi attack, with the host plant in the process of evolving mechanisms to stop the pathogen's advance [26], some studies show that some bryophyte species do not have their reproduction affected by parasitic fungi [18]. Also, associations between bryophytes and fungi can function in different ways, such as mycorrhizal, parasitic, and commensal [1]. Furthermore, they can have several benefits, both for fungi and for bryophytes, such as improvement in obtaining and cycling nitrogen, carbon (and other nutrients) and in the maintenance of more humid and protected environments [3, 4, 9, 34]. Many species of fungi associated to bryophytes are also found also in extreme environments, such as Antarctica, for example, probably indicating a symbiotic more than parasitic relationship [31, 38].

However, in Brazil these relationships Bryophyte/Agaricomycetes have not been characterized and studied and it is not known whether the occurrence of bryophytes was casual or whether a fungal association really occurred. This demonstrate the importance of more studies in this area in Brazil, what will make it possible to elucidate the ecological and physiological nature of these associations, among other issues, such as the influence of the environment on these associations.

7. Conclusions

Relationships involving bryophilous Agaricomycetes fungi and mosses are not yet well known. These associations are being studied around the world by optical and scanning electron microscopy, *in vitro* culture, sequencing of DNA, and phylogenetics analysis. In Brazil, a total of 33 species from four orders of Agaricomycetes were reported on bryophytes. Although much has been discovered about these interactions around the world, in Brazil these relationships are still not well characterized, either for lack of cooperation between mycologists and bryologists or for other reasons. Most of the Agaricomycetes species reported growing with mosses in Brazil occur in environments with high humidity and high altitude. However, the real influence of the environment on these associations has not yet been identified. Thus, it emphasizes the need for further studies on the interactions between bryophytes and Agaricomycetes, making it possible to better understand their ecology and taxonomy.

Acknowledgements

We would like to thank to the team of LATAF (Laboratório de Taxonomia de Fungos), of Universidade Federal do Pampa (UNIPAMPA) for the support to this research. This study was financed in part by the Coordenação de Aperfeiçoamento de Pessoal de Nível Superior – Brasil (CAPES) – Finance Code 001.

Author details

Cassiane Furlan-Lopes, Alice Lemos Costa, Marines de Avila Heberle, Fernando Augusto Bertazzo-Silva, Ana Luiza Klotz-Neves, Kamille Rodrigues Ferraz, Guilherme Henrique Mueller, Marina de Souza Falcão, Jorge Renato Pinheiro Velloso and Jair Putzke*
Programa de Pós-Graduação em Ciências Biológicas, Universidade Federal do Pampa, São Gabriel, Brazil

*Address all correspondence to: jrputzkebr@yahoo.com

References

[1] Davey ML, Currah RS. Interactions between mosses (Bryophyta) and fungi. Botany. 2006;**84**(10):1509-1519

[2] Read DJ, Duckett JD, Francis R, Ligrone R, Russell A. Symbiotic fungal associations in "lower" land plants. Philosophical Transactions of the Royal Society of London Series B-Biological Sciences. 2000;**2000**(355):815-831. DOI: 10.1098/rstb.2000.0617

[3] Field KJ, Pressel S, Duckett JG, Rimington WR, Bidartondo MI. Symbiotic options for the conquest of land. Trends in Ecology and Evolution. 2015b;**30**:477-486. DOI: 10.1016/j.tree.2015.05.007

[4] Pressel S, Bidartondo MI, Field KJ, Duckett JG. Advances in understanding of mycorrhizal-like associations in bryophytes. Bryophyte Diversity and Evolution. 2021;**43**(1):284-306

[5] Redecker D, Kodner R, Graham LE. Glomalean fungi from the Ordovician. Science. 2000;**289**(5486): 1920-1921. DOI: 10.1126/science.289.5486.1920

[6] Chang Y, Desirò A, Na H, Sandor L, Lipzen A, Clum A, et al. Phylogenomics of Endogonaceae and evolution of mycorrhizas within Mucoromycota. New Phytologist. 2019;**222**:511-525. DOI: 10.1111/nph.15613

[7] Trappe JM. What is a mycorrhiza? In: Proceedings of the 4th European Symposium on Mycorrhizae; Granada. Spain: EC Report EUR 16728. 1996. pp. 3-9

[8] Döbbeler P. Microniches occupied by bryophilous ascomycetes. Nova Hedwigia. 2002;**75**(3-4):275-306. DOI: 10.1127/0029-5035/2002/0075-0275

[9] Field KJ, Rimington WR, Bidartondo MI, Allinson KE, Beerling DJ, Cameron DD, et al. Functional analysis of liverworts in dual symbiosis with Glomeromycota and Mucoromycotina fungi under a simulated Palaeozoic CO2 decline. ISME Journal. 2016;**10**: 1514-1526. DOI: 10.1038/ismej.2015.204

[10] Döbbeler P, Hertel H. Bryophilous ascomycetes everywhere: Distribution maps of selected species on liverworts, mosses and Polytrichaceae. Herz. 2013; **26**(2):361-404. DOI: 10.13158/heia.26.2.2013.361

[11] Greiff GR, Cannon PF. Bryostroma popei—A new bryophilous ascomycete from the British Isles, with a worldwide key to the genus *Bryostroma*. Kew Bulletin. 2020;**75**(4):1-5. DOI: 10.1007/s12225-020-09917-0

[12] He MQ, Zhao RL, Hyde KD, Begerow D, Kemler M, Yurkov A, et al. Notes, outline and divergence times of Basidiomycota. Fungal diversity. 2019; **99**(1):105-367. DOI: 10.1007/s13225-019-00435-4

[13] Prance GT. The use of edible fungi by Amazonian Indians. Advances in Economic Botany. 1984;**1**:127-139

[14] Kumar Sharma S, Gautam N. Chemical and bioactive profiling, and biological activities of coral fungi from Northwestern Himalayas. Scientific Reports. 2017;**7**(1):1-13

[15] Ajith TA, Janardhanan KK. Cytotoxic and antitumor activities of a polypore macrofungus, *Phellinus rimosus* (Berk) Pilat. Journal of Ethnopharmacology. 2003;**84**:157-162

[16] Vital DM, Capelari M, Gugliotta AM, Bodoni VLR. Bryophytes on fungi. Tropical Bryology. 2000;**19**:31-40

[17] Kauserud H, Mathiesen C, Ohlson M. High diversity of fungi associated with living parts of boreal forest bryophytes. Botany. 2008;**86**(11): 1326-1333

[18] Korotkin HB, Swenie RA, Miettinen O, Budke JM, Chen KH, Lutzoni F, et al. Stable isotope analyses reveal previously unknown trophic mode diversity in the Hymenochaetales. American Journal of Botany. 2018; **105**(11):1869-1887

[19] Davey ML, Heegaard E, Halvorsen R, Kauserud H, Ohlson M. Amplicon-pyrosequencing-based detection of compositional shifts in bryophyte-associated fungal communities along an elevation gradient. Molecular Ecology. 2013;**22**(2): 368-383

[20] Singer R. Type Studies on Basidiomycetes VI. Lilloa. 1953a;**26**: 57-159

[21] Albuquerque HR, Araújo JPM, Putzke J. *Chromocyphella muscicola* (Fr.) Donk. (Basidiomycota, Agaricales): primeira citação para o Brasil. Revista Brasileira de Biociências. 2007;**5**: 999-1001

[22] Lücking R, Leavitt SD, Hawksworth DL. Species in lichen-forming fungi: balancing between conceptual and practical considerations, and between phenotype and phylogenomics. Fungal Diversity. 2021; **109**(1):99-154. DOI: 10.1007/s13225-021-00477-7

[23] Wilk K, Pabijan M, Saługa M, Gaya E, Lücking R. Phylogenetic revision of South American Teloschistaceae (lichenized Ascomycota, Teloschistales) reveals three new genera and species. Mycologia. 2021;**113**(2): 278-299

[24] Grandi RAP, Silva PD, Vital DM. Hyphomycetes (fungos conidiais) associados a briófitas em decomposição. Acta Botanica Brasilica. 2008;**22**(2): 599-606. DOI: 10.1590/S0102-33062008000200027

[25] Putzke J, Putzke MTL. Cogumelos-Fungos agaricales no Brasil, familias Agaricaceae, Amanitaceae, Bolbitaceae, Entolomataceae, Coprinaceae/Psathyrellaceae, Crepidotaceae e Hygrophoraceae. 1st ed. São Gabriel: Editora JP; 2017. p. 518

[26] Redhead SA. Parasitism of bryophytes by agarics. Canadian Journal of Botany. 1981;**59**:63-67. DOI: 10.1139/b81-011

[27] Döbbeler P. Ascomycetes growing on *Polytrichum sexangulare*. In: Laursen GA, Ammirati JF, Redhead SA, editors. Arctic and Alpine Mycology II. Environmental Science Research. 34th ed. Boston: Springer; 1987. DOI: 10.1007/978-1-4757-1939-0_8

[28] Hein JD, Bart FVT. Bryophyte interactions with other plants. Botanical Journal of the Linnean Society. 1990;**104** (1–3):79-98. DOI: 10.1111/j.1095-8339.1990.tb02212.x

[29] Bujakiewicz A. Macrofungi on soil in deciduous forests. In: Winterhoff W, editor. Fungi in Vegetation Science. Handbook of Vegetation Science. 19th ed. Dordrecht: Springer; 1992. pp. 49-78. DOI: 10.1007/978-94-011-2414-0_3

[30] Force TET. Standardized Inventory Methodologies for Components of British Columbia's Biodiversity. 1st ed. British Columbia: Resources Inventory Committee; 1997. p. 57

[31] Smith RIL. Species-diversity and resource relationships of South Georgian fungi. Antarctic Science. 1994;**6**:45-52

[32] Di Battista C, Selosse MA, Bouchard D, Stenström E, Le Tacon F. Variations in symbiotic efficiency, phenotypic characters and ploidy level among different isolates of the ectomycorrhizal basidiomycete *Laccaria bicolor* strain S 238. Mycological Research. 1996;**100**(11):1315-1324

[33] Arnolds E. The analysis and classification of fungal communities with special reference to macrofungi. In: Winterhoff W, editor. Fungi in Vegetation Science. Handbook of Vegetation Science. 19th ed. Dordrecht, Springer. pp. 7-47. DOI: 10.1007/978-94-011-2414-0_2

[34] Bowden RD. Inputs, outputs, and accumulation of nitrogen in an early successional moss (Polytrichum) ecosystem. Ecological Monographs. 1991;**61**(2):207-223

[35] Moreno G, Contu M, Ortega A, Platas G, Peláez F. Molecular phylogenetic studies show *Omphalina giovanellae* represents a new section of Clitopilus (Agaricomycetes). Mycological Research. 2007;**111**(12):1399-1405

[36] Nelsen MP, Lücking R, Umaña L, Trest MT, Will-Wolf S, Chaves JL, et al. *Multiclavula ichthyiformis* (Fungi: Basidiomycota: Cantharellales: Clavulinaceae), a remarkable new basidiolichen from Costa Rica. American Journal of Botany. 2007;**94**(8): 1289-1296

[37] Thormann MN, Rice AV. Fungi from peatlands. Fungal Diversity. 2007; **24**(2415):299

[38] Ludley KE, Robinson CH. Decomposer basidiomycota in Arctic and Antarctic ecosystems. Soil Biology and Biochemistry. 2008;**40**(1):11-29

[39] Uzun Y, Karacan İH, Yakar S. New additions to Turkish agaricales. Biological Diversity and Conservation. 2008;**10**(2):8-13

[40] Dam N, Boomsluiter M. Fairy rings in moss. Coolia. 2009;**52**(2):67-72

[41] Bodensteiner P. *Maireina afibulata* and *M. attenuatipilis*, new members of the cyphelloid genus *Maireina* (Basidiomycota, Agaricomycetes). Mycological Progress. 2007;**6**(4):221-228. DOI: 10.1007/s11557-007-0538-9

[42] Lawrey JD, Lücking R, Sipman HJ, Chaves JL, Redhead SA, Bungartz F, et al. High concentration of basidiolichens in a single family of agaricoid mushrooms (Basidiomycota: Agaricales: Hygrophoraceae). Mycological Research. 2009;**113**(10): 1154-1171

[43] Ronikier A, Ronikier M. Biogeographical patterns of arctic-alpine fungi: distribution analysis of *Marasmius epidryas*, a typical circumpolar species of cold environments. North American Fungi. 2010;**5**(5):23-50. DOI: 10.2509/naf2010.005.0053

[44] Prylutsky OV. Agarics and Boleti (Agaricomycetes, Basidiomycota) of Sphagnum peatbogs and swampy forests of "Slobozhansky" National Natural Park. Studia Biologica. 2014;**8**(3–4): 179-186

[45] Singer R. The Agaricales in Modern Taxonomy. 4th ed. Germany: Koeltz Scientific Books; 1986. p. 981 88pl

[46] Putzke J, Putzke MTL. Cogumelos (fungos Agaricales) no Brasil, Família Tricholomataceae. 1st ed. Vol. III. São Gabriel, Editora JP; 2022 in press. p. 356

[47] Kaya A, Akata I, Uzun Y. Two new records for Turkish Agaricales. Biological Diversity and Conservation. 2013;**6**(3): 150-152

[48] GBIF.org. GBFI occurrence download [internet] Available from: DOI: 10.15468/dl.djk3kh. [Accessed: 2022-06-26]

[49] Nita J, Bujakiewicz A. Riparian and alder forests – shelters for rare and threatened macromycetes. In: Mirek Z, Nikel A, editors. Rare, Relict and Endangered Plant and Fungi in Poland. Szafer Institute of Botany, Polish Academy of Sciences: Kraków; 2009. pp. 325-334

[50] De Oliveira LA, de Jesus MA, Matsuura ABJ, Gasparotto L, Oliveira JDS, de Lima-Neto RG, et al. Conhecimento, Conservação e Uso de Fungos. 1st ed. Manaus: Editora Inpa; 2019. p. 224

[51] Furtado AN, Daniels PP, Neves MA. New species and new records of Clavariaceae (Agaricales) from Brazil. Phytotaxa. 2016;**253**(1):1-26. DOI: 10.11646/phytotaxa.253.1.1

[52] Gomes DCV, de Alencar MVOB, Dos Reis AC, de Lima RMT, de Oliveira Santos JV, da Mata AMOF, et al. Antioxidant, anti-inflammatory and cytotoxic/antitumoral bioactives from the phylum Basidiomycota and their possible mechanisms of action. Biomedicine & Pharmacotherapy. 2019;**112**:108643. DOI: 10.1016/j.biopha.2019.108643

[53] Hawkeswood TJ, Sommung B, Sommung A. First record of the coral fungus, Clavaria fragilis Holmsk: Fr. (1790) (Basidiomycota: Clavariaceae) from Sisaket Province, Thailand. Calodema. 2020;**820**:1-3

[54] Burt EA. The North American species of *Clavaria* with illustration of the type specimens. Annals of the Missouri Botanical Garden. 1922;**9**:1-78

[55] Corner EJH. Notes on *Clavaria*. Transactions of the British Mycological Society. 1967;**50**:33-44

[56] Singer R. Mycoflora australis. Nova Hedwigia. 1969;**29**(1):1-405

[57] De Meijer AAR. Notable macrofungi from Brazil's Paraná pine forests. 1st ed. Curitiba: Embrapa Florestas; 2008. p. 418

[58] Putzke J, Putzke MTL. Cogumelos (fungos Agaricales) no Brasil, Ordens Boletales (Boletaceae e Paxillaceae), Polyporales (Polyporaceae/Lentinaceae), Russulales (Russulaceae) e Agaricales (Cortinariaceae, Inocybaceae, Pluteaceae e Strophariaceae). 1st ed. Vol. II. São Gabriel: Editora JP; 2018. p. 380

[59] Tănase C, Pop A. Red list of Romanian macrofungi species. In: Bioplatform–Romanian National Platform for Biodiversity. Bucureşti: Edit. Academiei Române; 2005. pp. 101-107

[60] Grzesiak B, Wolski GJ. Bryophilous species of the genus *Galerina* in peat bogs of central Poland. Herz. 2015;**28**(2): 607-623. DOI: 10.13158/heia.28.2.2015.607

[61] Horak E. Caterina (Agaricales) in neotropical South America. Type studies, additional material, comments, key. Boletín de la Sociedad Argentina de Botánica. 1992;**28**:233-246

[62] Doyle MF. A floristic survey of the fleshy fungi from the Hawaiian Islands with notes on common marasmioid fungi from southern Illinois. [thesis], Illinois: Southern Illinois University. 1985.

[63] Guzmán G. The genus *Psilocybe* a systematic revision of the known species including the history, distribution and chemistry of the hallucinogenic species Beihefte. Nova Hedwigia. 1983;**74**:1-439

[64] Coimbra VRM, Gibertoni TB. First record of *Trichopilus fasciculatus* (Agaricales) from Brazil, with a key for

the species of Entolomataceae from the Northern region. Mycoscience. 2015; **56**(1):118-122. DOI: 10.1016/j.myc.2014.04.004

[65] Wang K, Zhao M, Su J, Yang L, Deng H, Wang Y, et al. The use of Checklist of Fungi in China database in the red list assessment of macrofungi in China. Biodiversity Science. 2020;**28**(1): 74. DOI: 10.17520/biods.2019163

[66] Stamets P. Psilocybin Mushrooms of the World. Berkeley: Ten Speed Press; 1996. p. 229

[67] Strauss D, Ghosh S, Murray Z, Gryzenhout M. An Overview on the Taxonomy, Phylogenetics and Ecology of the Psychedelic Genera *Psilocybe*, *Panaeolus, Pluteus* and *Gymnopilus*. Frontiers in Forests and Global Change. **5**:813998. DOI: 10.3389/ffgc.2022.813998

[68] Pegler DN. Agaric Flora of the Lesser Antilles. Kew Bulletin; 1983. p. 668

[69] Dal-Forno M, Lücking R, Bungartz F, Yánez-Ayabaca A, Marcelli MP, Spielmann AA, et al. From one to six: Unrecognized species diversity in the genus *Acantholichen* (lichenized Basidiomycota: Hygrophoraceae). Mycologia. 2016; **108**(1):38-55. DOI: 10.3852/15-060

[70] Liu YP, Dai Q, Wang WX, He J, Li ZH, Feng T, et al. Psathyrins: antibacterial diterpenoids from *Psathyrella candolleana*. Journal of Natural Products. 2020;**83**(5): 1725-1729. DOI: 10.1021/acs.jnatprod.0c00269

[71] Da Silva PS, Cortez VG, da Silveira RMB. The mycobiota of Itapuã Park, Rio Grande do Sul, Brazil. I. Species of Strophariaceae (Agaricales). Mycotaxon-Ithaca. 2006;**97**:219

[72] Cortez VG, Silveira RMBD. Species of *Hypholoma* (Fr.) P. Kumm. (Strophariaceae, Agaricales) in Rio Grande do Sul State, Brazil. Acta Botanica Brasilica. 2007;**21**:609-621

[73] Ćetković I, Tkalčec Z, Dragićević S, Alegro A, Šegota V, Jadan M, et al. *Entoloma conferendum*, *Hygrocybe coccineocrenata*, and *Hypholoma ericaeum* new to Montenegro. Mycotaxon. 2020; **135**(3):637-647

[74] Kaygusuz O, Şengül MŞ. *Callistosporium luteo-olivaceum* (Callistosporiaceae: Basidiomycota), an agaric fungi new to Turkey. Anatolian Journal of Botany. 2021;**6**(1):7-12. DOI: 10.30616/ajb.1014854

[75] An GH, Han JG, Cho JH. Antioxidant activities and β-glucan contents of wild mushrooms in Korea. Journal of Mushroom. 2019;**17**(3):144-151. DOI: 10.14480/JM.2019.17.3.144

[76] Martínez M, Niveiro N. *Gymnopus aquosus* (Bull.) Antonín & Noordel. (Agaricales–Omphalotaceae): Nuevo registro para la micobiota de Paraguay. Steviana. 2019;**11**(2):41-47

[77] Raithelhuber J. Flora Mycologica Argentina: Hongos III. Stuttgart: Mycosur; 1991

[78] Ge Y, Liu Z, Zeng H, Cheng X, Na Q. Updated description of Atheniella (Mycenaceae, Agaricales), including three new species with brightly coloured pilei from Yunnan Province, southwest China. MycoKeys. 2021;**14**(81):139-164. DOI: 10.3897/mycokeys.81.67773

[79] Souza HQD, Aguiar IDJA. Diversidade de Agaricales (Basidiomycota) na Reserva Biológica Walter Egler, Amazonas, Brasil. Acta Amazonica. 2004;**34**:43-51. DOI: 10.1590/S0044-59672004000100006

[80] Magnago AC, Neves MA. New record of *Austroboletus festivus* (Boletaceae) from Santa Catarina, Brazil. Brazilian Journal of Botany. 2014;**37**: 197-200. DOI: 10.1007/s40415-014-0048-3

[81] Singer R. Type studies in Basidiomycetes. X. Persoonia-Molecular Phylogeny and Evolution of Fungi. 1961; **2**(1):1-62

[82] Watling R, De Meijer AR. Macromycetes from the state of Paraná, Brazil: 5. Poroid and lamellate boletes. Edinburgh Journal of Botany. 1997; **54**(2):231-251

[83] Fulgenzi TD, Halling RE, Henkel TW. *Fistulinella cinereoalba* sp. nov. and new distribution records for Austroboletus from Guyana. Mycologia. 2010;**102**(1):224-232

[84] Gibertoni TB, Gomes-Silva AC, Chikowski RS, Lira CRS, Soares AMS, Melo GSN, et al. Hymenochaetales in Lista de Espécies da Flora do Brasil. Jardim Botânico do Rio de Janeiro. [internet]. 2015 Available from: http://floradobrasil2015.jbrj.gov.br/FB92820. [Accessed: 2022-07-15]

[85] Alves-Silva G, Reck MA, da Silveira RMB, Bittencourt F, Robledo GL, Góes-Neto A, et al. The Neotropical Fomitiporia (Hymenochaetales, Basidiomycota): the redefinition of *F. apiahyna* ss allows revealing a high hidden species diversity. Mycological Progress. 2020;**19**(8): 769-790. DOI: 10.1007/s11557-020-01593-5

[86] Laovachirasuwan P, Judprakob C, Sinaphet B, Phadungkit M. *In vitro* antioxidant and antimutagenic activities of different solvent extracts of *Phellinus* spp. International Food Research Journal. 2016;**23**(6):2608-2615

[87] Gibertoni TB, Gomes-Silva AC, Chikowski RS, Lira CRS, Soares AMS, Melo GSN, et al. Hymenochaetales in Lista de Espécies da Flora do Brasil. Jardim Botânico do Rio de Janeiro. [internet]. 2015. Available from: http://floradobrasil2015.jbrj.gov.br/jabot/floradobrasil/FB95281 [Accessed: 2022-07-15]

[88] Silva FS, de Sá MS, Costa JFO, Pinto FP, Lima MS, Lucchese AM, et al. *In vitro* pharmacological screening of macrofungi extracts from the Brazilian northeastern region. Pharmaceutical Biology. 2009;**47**(5):384-389. DOI: 10.1080/13880200902755226

[89] Drechsler-Santos ER, Santos PJP, Gibertoni TB, Cavalcanti MAQ. Ecological aspects of Hymenochaetaceae in an area of Caatinga (semi-arid) in Northeast Brazil. Fungal Diversity. 2010; **42**(1):71-78

[90] De Campos-Santana M, Robledo G, Decock C, Da Silveira RMB. Diversity of the Poroid Hymenochaetaceae (Basidiomycota) from the Atlantic Forest and Pampa in Southern Brazil. Cryptogamie, Mycologie. 2015;**36**(1): 43-78

[91] Da Silva MA. Diversidade da família hymenochaetaceae donk da amazônia brasileira e avaliação da bioatividade de extratos de phellinus crystallis, uma nova espécie com perspectiva futura de potencial biotecnológico [thesis]. Manaus: Instituto Nacional de Pesquisas da Amazônia; 2021

[92] Chang ZQ, Gebru E, Lee SP, Rhee MH, Kim JC, Cheng H, et al. In vitro antioxidant and anti-inflammatory activities of protocatechualdehyde isolated from *Phellinus gilvus*. Journal of Nutritional Science and Vitaminology. 2011;**57**(1):118-122

[93] Huo J, Zhong S, Du X, Cao Y, Wang W, Sun Y, et al. Whole-genome sequence of *Phellinus gilvus* (mulberry Sanghuang) reveals its unique medicinal values. Journal of Advanced Research. 2020;**24**:325-335

[94] Abrahão MC, Gugliotta ADM, Gomes E. Poliporóides (Basidiomycota) em fragmentos de mata no perímetro urbano de São José do Rio Preto, São Paulo, Brasil. Brazilian Journal of Botany. 2009;**32**:427-440

[95] Gibertoni TB, Gomes-Silva AC, Chikowski RS, Lira CRS, Soares AMS, Melo GSN, et al. Hymenochaetales in Lista de Espécies da Flora do Brasil. Jardim Botânico do Rio de Janeiro. [internet. 2015. Available from: http://floradobrasil2015.jbrj.gov.br/jabot/floradobrasil//FB92375 Accessed: 2022-06-16

[96] Pagin-Cláudio F, Gugliotta ADM, Pires RM. Checklist of polypores (Agaricomycetes, Basidiomycota) reveals seventeen new records in Brazilian Atlantic Forest fragments at the Poços de Caldas Plateau, Minas Gerais. Rodriguésia. 2022;**73**:e00832021. DOI: 10.1590/2175-7860202273058

[97] Funez LA, Monteiro M, Costa-Rezende DH, Alves-Silva G, Drechsler-Santos ER. *Rickenella fibula* (Repetobasidiaceae:Basidiomycota): a tiny species with wide distribution also occurs in Brazil. Iheringia Serie Botanica. in press. 2022

[98] Stijve T, Kuyper TW. Absence of psilocybin in species of fungi previously reported to contain psilocybin and related tryptamine derivatives. Persoonia. 1988;**13**(4):463-465

[99] Gugliotta AM, Gibertoni TB, Drechsler-Santos ER, Silveira RMB, Chikowski RS, Pires RM, et al. Polyporales in Lista de Espécies da Flora do Brasil. Jardim Botânico do Rio de Janeiro. [internet]. 2015. Available from: http://floradobrasil2015.jbrj.gov.br/jabot/floradobrasil/FB95281 [Accessed: 2022-06-15]

[100] Gugliotta AM, Bononi VLR. Polyporaceae do Parque Estadual da Ilha do Cardoso, São Paulo, Brasil. Vol. 12. Boletim do Instituto de Botânica; 1999. pp. 1-112

[101] Gugliotta AM, Gibertoni TB, Drechsler-Santos ER, Silveira RMB, Chikowski RS, Pires RM, et al. Polyporales in Lista de Espécies da Flora do Brasil. Jardim Botânico do Rio de Janeiro. [internet. 2015. Available from: http://floradobrasil2015.jbrj.gov.br/FB92447 Accessed: 2022-06-15

[102] Costa DP, Peralta DF. Bryophytes diversity in Brazil. Rodriguésia. 2015;**66**:1063-1071

[103] Dewes TDS, Santos NDD, Bordin J. What does a phytophysiognomic mosaic reveal about mosses and liverworts from the subtropical Atlantic Forest? Acta Botanica Brasilica. 2021;**35**:517-531

Chapter 5

Why Are Moss Biocrusts Necessary for System Conservation in a Semiarid Region of Southern Argentina?

Alicia Kröpfl

Abstract

The importance of biological crusts in arid and semiarid ecosystems has been widely recognized in the last decades, but their function is still not fully known, much less so in our country (Argentina) and region, where they have often gone unnoticed. Biological crusts appear in sites with a low level of disturbance and have a fundamental role in maintaining the soil surface structure. In the Monte ecoregion of Rio Negro province, Argentina, this layer is dominated by moss, and it is present in diverse physiognomic-floristic types of vegetation, but it tends to disappear in sites disturbed by grazing and by severe fires, and at sites where clearing methods were used to remove the soil surface. The multiple benefits that these crusts can provide to ecosystems justify the need to intensify the knowledge of their structure and functioning, to understand the particular role that they fulfill and to be able to manage these systems by taking this component into account.

Keywords: disturbances, water balance, seed bank, seedling emergence

1. Introduction

The vegetation that characterizes the southern Monte ecoregion is a shrubby steppe, and, as in other drylands, it is scarce and arranged in islands of vascular plants with large interspaces between them, covered to a greater or lesser extent by a herbaceous layer. The third functional group that accompanies the shrubs and the herbaceous layer is that of the biological crusts, which is globally known as a diverse soil surface community of cyanobacteria, algae, fungi, lichens, and bryophytes [1]. Research on these biological crusts has received considerable attention, especially since Belnap and Lange´s publication [2], so that knowledge about them has grown exponentially throughout the world, highlighting their sensitivity to global change. However, in South America, there are still gaps in the framework of that knowledge [3].

It has been widely accepted that succession in biological crusts follows a general pattern, starting with cyanobacteria and algae and concluding with bryophytes at the later successional stages, probably due to their greater hygroscopicity, higher growth rates, relative height, and deeper rhizoids [4, 5]. However, recent studies have proposed that mosses can be present from the initial phases of succession if the conditions are favorable [2, 6].

Several studies in recent decades have shown that this component of dryland ecosystems [7] is often not taken into account but has many important functions in the sustainability of these systems, such as aggregation of soil particles (resulting in soil protection), seed retention and germination [8], water infiltration [9], reduction of wind and water erosion [10], as well as nitrogen fixation [11–13], and carbon sequestration [14]. In addition, if a disturbed site is given time to re-establish the crust, it would also improve the results in terms of the establishment of seedlings [1]. Its role is especially important to maintain the stability of the soil surface against the impact of raindrops in those soils which, due to their physical and chemical characteristics, tend to form vesicular surface crusts [15] and favor desertification processes in the face of successive wetting and drying cycles. All these attributes allow us to designate biocrusts as "ecosystem engineers."

In general, the effect of disturbances has been studied on the two most visible guilds of the Monte ecoregion (grasses and shrubs), although a rational management of natural resources should consider all its components, taking account of their function within the system. The biological crust has a fundamental role not only in the conservation of the superficial structure of the soil and the possibilities of regeneration of the herbaceous cover, but also in its contribution to biodiversity. In a States and Transitions model that we proposed [16], we were able to establish that the original system in the place where we carry out our studies was formed by two states that integrated a single domain of attraction and had a high resilience. The reduction of the herbaceous and biological crust layers, and the changes in the superficial structure of the soil, generated a distance from this domain of attraction, leading the system to other very stable states of lower productivity, and would be responsible for the advance of the desertification process.

Biological crusts in the eastern Monte ecoregion are dominated by moss, and their main species are *Syntrichia prínceps* (De Not.) Mitt. and *Ceratodon purpureus* (Hedw.) (**Figure 1**) [17].

Assuming that the presence of biological crusts would be associated with the lack of disturbances which is verified in more stable systems, some authors are using the presence of biological crusts as an indicator of the condition or "health" of grasslands [18–20]. In the same sense, Song et al. [21] concluded that biological crusts act as natural regulators for vegetation patterns and thus promote ecosystem stability and sustainability.

Concerning climate change, perhaps the most worrying and great environmental problem today, studies by Rutherford et al. [22] utilizing climate manipulation treatments suggest that the elimination of key species of mosses and lichens from the biological crust community may have dramatic effects on the biogeochemical and hydrological functions in drylands.

Also, the reduction in biocrust cover due to warming will lessen the capacity of drylands to sequester atmospheric CO_2. This decrease may act synergistically with other warming-induced effects, altering C cycling in drylands, and reducing soil C stocks in the mid to long term [23], which is one of the most globally valued functions in ecosystems today. In accordance with this, Durán et al. [24] propose the use of the specialized microbiome of biocrusts to be applied in a new environment to counteract the negative effects of climate change.

Figure 1.
Moss biocrust with Syntrichia princeps (a) and Ceratodon purpureus (b) species.

Our concern in recent years has been trying to highlight the presence of moss in our ecosystem and to quantify some of its functions and how it is affected by the occurrence of disturbances at different scales.

2. Moss biocrusts functions

2.1 Soil protection

In the eastern Monte region, moss biocrusts cover a variety of soil surfaces, according to the physiognomic type of vegetation and the degree of disturbance affecting it, although these biocrusts are highly vulnerable to alterations by both natural and anthropogenic disturbances [25], and among these, domestic livestock trampling is recognized as the most widespread [26].

The trampling of grazing animals has a negative effect on arid and semiarid grasslands by causing fragmentation of the biological crust, and also, the loss of this crust and the herbaceous cover, contributing to an increase in bare soil, facilitating the processes of erosion, loss of nutrients, and the formation of vesicular crusts, which are difficult to reverse [27].

In addition, chaining used to remove the shrub layer increases the area that can be trampled by livestock, potentially generating an indirect negative effect on these crusts.

We studied the effect of some of those disturbances on the cover of moss biocrusts in relation to intact sites, and, as these organisms are of small size, it must be taken into account that microenvironments are often determining factors in their maintenance within those systems.

As described earlier, the vegetation of our region is grouped into islands with interspaces between them, which have different microenvironmental characteristics:

	Grazed	Ungrazed	*p*
Flat sites	19.90	46.89	<0.01
Mounds	32.13	38.98	ns
Total	26.49	43.87	<0.05

Table 1.
Mean cover of moss biocrust (%) in grazed and ungrazed areas, grouped according to their microtopographic location.

the islands constitute phytogenic mounds, where shrubs offer shade, protection, and nutrients to the vegetation growing underneath them, and the interspaces are flat sites more or less vegetated, with opposite environmental conditions (wind exposure, run-off, trampling access, and increased insolation). When we analyzed the differences in the cover of moss biocrust between both kind of sites under grazed and ungrazed conditions, we found a significant reduction in the flat sites between the shrubs, where trampling had an undeniable presence (**Table 1**). The lack of such difference in the mounds is strongly conditioned by the difficulty of access that shrubs impose on cattle, even though the whole field is being grazed.

The results of Yang et al. [14] showed that, in addition to the direct damage to the moss, there is a significant change in the environment that will affect it, because trampling disturbance increases carbon emissions from biocrust soils. These losses of CO_2 from biocrust soils after disturbance, in turn, may substantially reduce the biocrust contribution to the soil carbon budget.

We also found that, if a site was mechanically disturbed by chaining, in addition to being grazed, the cover was significantly reduced by about 33%, and, when we compared ungrazed and grazed conditions in a chained site, the reduction was of 36%, as we had found in another study [17].

In a previous trial where we compared the effect of clearing, fire and both disturbances occurring successively on the same site, there was a significant decrease in the moss cover with any of the disturbances (**Figure 2**), thus leaving more bare ground than in the control site [28] and confirming the general theory.

However, shrubs not only provide protection against trampling, but also they offer shady conditions which improve conditions for the development and survival of moss. On grazed sites, shading is due to the shrubs and other vascular plants that grow underneath them, but on ungrazed sites it is the grasses and herbs that form the intact herbaceous layer that mainly provides shading.

When we compared moss cover between sunny and shady sites in a field with small shrubs (as it had been cleared 3 years previously), we always found differences in favor of shaded sites, regardless of whether or not the environment was grazed (**Table 2**).

Although, when we compared the sunny and shady sites with each other, we did not find any significant differences, suggesting that the degree of sunlight was more important than the location of moss in the environment and which functional group provided the shade. However, if the shading is too intense, the moss biocrust would be affected because the relationship goes from facilitation to competition: according to Zhang et al. [29], vascular plant communities can affect biocrust development, composition, and function through canopy shading, although a dense canopy can deprive crusts of adequate light for photosynthesis.

To analyze the effect of mechanical removal, we carried out a trial where we compared the soil cover at a site with complete removal of the shrubs (clearing) and

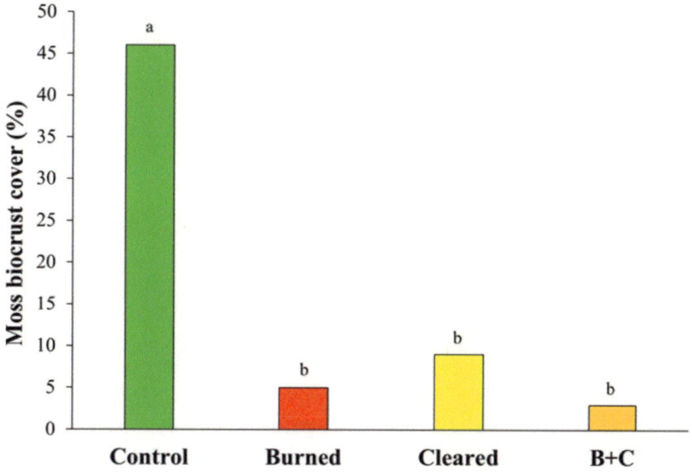

Figure 2.
Moss biocrust cover (%) in three disturbed sites (burned, cleared, and burned + cleared) and a control site. Lowercase letters indicate significant differences ($p < 0.05$).

		Ungrazed		Grazed	
Site		$\dot{x} \pm SE$	p	$\dot{x} \pm SE$	p
M	Sun	18.1 ± 9.8	0.05	21.2 ± 4.0	0.013
	Shadow	39.2 ± 5.4		42.4 ± 4.7	
F	Sun	28.3 ± 4.1	0.01	19.4 ± 3.1	0.025
	Shadow	53.1 ± 6.2		34.6 ± 6.3	

Table 2.
*Moss biocrust cover (%) in sunny or shady locations of mounds (M) and flat (F) sites, ungrazed or grazed in the same field, with the level of significance of the differences between them (**p**).*

a partial one (thinning) with an intact site (Monte), in a closed area and another area grazed for 3 years. Every year we found the same situation repeated: the intact site significantly outperformed the modified ones, and the thinned site outperformed the cleared one, showing a graduation of damage as the level of shrub removal (small-scale disturbance) increased. There were no differences between years ($p = 0.16$), or between grazed and ungrazed sites ($p = 0.078$) (**Figure 3**).

In addition to trampling and mechanical removal, another disturbance of great magnitude on the cover of moss biocrust is fire, although the damage will depend on the severity of the fire. *Ceratodon purpureus*, in particular, can resist high temperatures and apparently can resume its activity when humidity conditions allow, as we have noticed in the field. After a fire season in the region, Bran et al. [30] analyzed the effects of fire severity on the vegetation, classifying sites with a burn severity index (SI) established visually into categories from 0 (unburned control) to 5 (maximum severity) depending on the remaining standing biomass of the shrub layer. In unpublished data from that paper, Cecchi (*pers. com.*) found a strong subsequent reduction in moss cover in relation to the degree of severity of the fire (**Figure 4**). It can be seen that, from severity level 3 ("shrubs with most of their structure standing, branches of less than 0.6 cm in diameter conserved, leaves

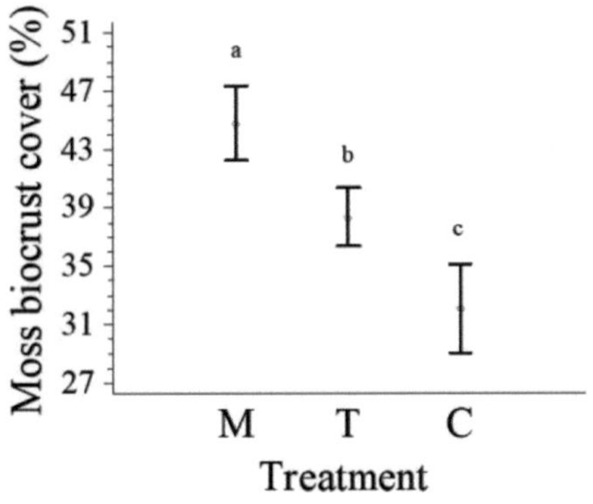

Figure 3.
Moss biocrust cover (%) in three vegetation situations, control (M), shrub clearing (C), and thinning (T). Lowercase letters indicate significant differences ($p < 0.01$).

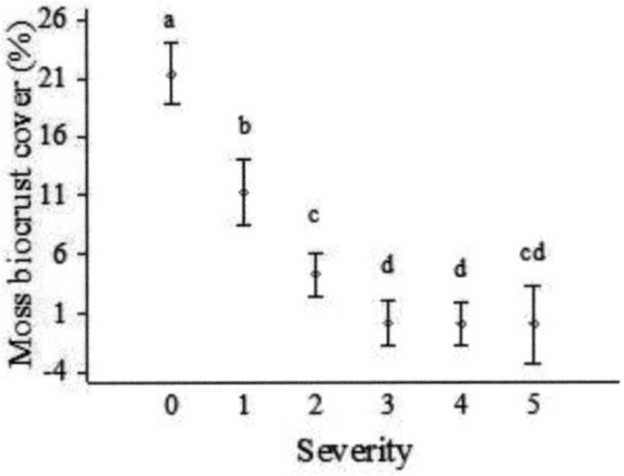

Figure 4.
Moss biocrust cover (%) in relation to increasing degree of burn severity. Lowercase letters indicate significant differences ($p < 0.05$). Source: Gustavo Cecchi.

totally destroyed by fire"), the damage would be the same and the moss coverage does not exceed 1%.

If the fire is severe, one of the results it produces is a redistribution of the sediments accumulated underneath the shrubs, homogenizing the soil surface, modifying the soil texture in the interspaces, and eliminating the sheltered sites provided by the shrubs and their nursing effect. This means that the moss biocrust will take some time to recover, and quantitative studies are still lacking in our region to be able to estimate how long that time could be. In this sense, Brianne et al. [31] performed a meta-analysis to gather information about the recovery of biocrust cover following fire at the global scale and highlighted the need to advance with that research across a broader geographic range.

Novel approaches for *ex situ* cultivation and inoculation are now being developed for using these communities in large-scale post-fire ecosystems restoration [32], thereby considering moss biocrust as a tool for accelerating soil restoration in semiarid ecosystems affected by wildfires [33].

2.2 Water balance

Mosses can use the surface moisture of the soil that appears in pulses in these systems and have rapid physiological responses [34], which would favor their primary productivity. In fact, they only need a short-wet time to recover their metabolic activity after a drought period and maybe this would allow them to be considered as "resurrection plants."

Positive influences of biological crusts on surface hydrological processes in semi-arid ecosystems have been documented by many authors [9, 35–37]. The roughness of moss biocrusts creates a larger surface for the detention of rainwater, so it contributs to increased infiltration in those sites where mosses are the main component of the biological crusts [5]. This can also reduce runoff and the risk of water erosion in sloping areas, since it decreases the kinetic energy of the water and therefore its erosive force [38].

Since the observations of St. Clair et al. [39], it has been known that the loss of moss biocrust of the soils surface due to disturbances can affect infiltration.

To evaluate water content in the field, we extracted superficial cores in sites with and without moss biocrust and weighed them following the gravimetric method, after removing moss present of the surface. We found significant differences in water storage between both the two kinds of surfaces (**Figure 5**).

Also, to evaluate the effect of the moss biocrust on the water storage capacity and on water evaporation more accurately, we extracted undisturbed soil cores from an exclosure in spring, either with or without moss biocrust, placed them in pots to transfer them to a greenhouse, and we repeated this trial in autumn [17]. Water storage capacity was significantly higher in the soil with moss biocrust than in the soil without mosses, for both dates ($p = 0.002$ and $p = 0.05$, respectively) (**Figure 6**).

In the first trial, we also calculated the daily loss of water by the difference in weight until no more evaporation occurred, and the water evaporation rate was lower with moss biocrust than without it. On that date, we also prepared the same number of samples extracting the moss layer (MB extracted), and we verified that the soil behaved in exactly the same way as that of the samples that preserved the moss

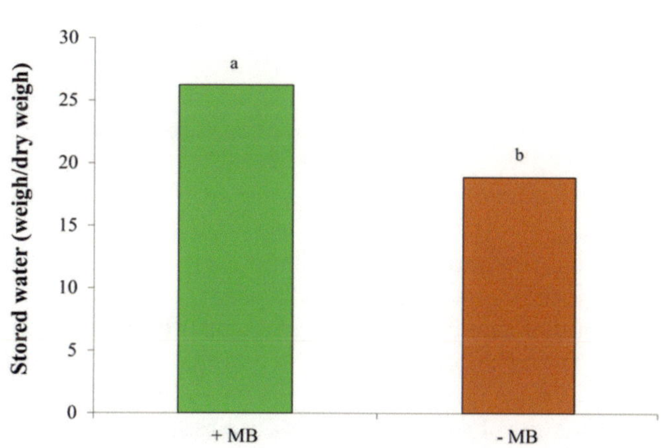

Figure 5.
Water content stored in samples with (+ MB) and without (– MB) moss biocrust. Lowercase letters indicate significant differences ($p < 0.01$).

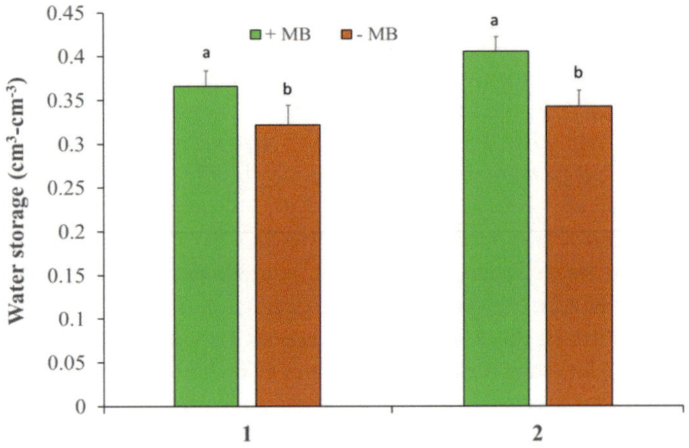

Figure 6.
Water storage (volume/volume) in samples with and without moss biocrust (+SE), for two sample dates (1: spring and 2: autumn). Lowercase letters indicate significant differences (p < 0.01).

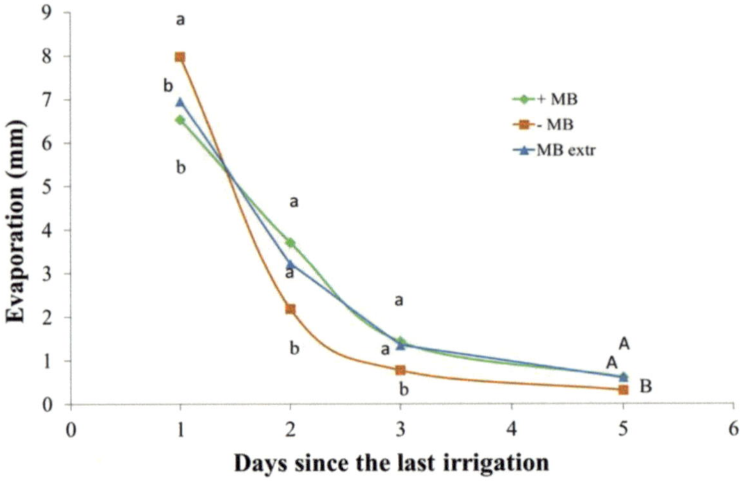

Figure 7.
Daily surface evaporation since the last irrigation for plots with (+MB), without (– MB) mosses biological crust, and with moss biocrust extracted (MB extr). Lowercase letters indicate significant differences (p < 0.01) and capital letters indicate differences at p < 0.1. Source: Gustavo Cecchi.

(**Figure 7**), showing that the water retention benefit provided by the mosses persisted initially, even when the mosses were no longer present.

Our results suggest that the sites with moss biocrusts have surface water storage conditions that could favor the germination and emergence of seedlings, maintaining the water balance as they have greater water availability and lower evaporation rates in the first centimeters of soil. This agrees with results found by Chamizo et al. [9], who concluded that biocrusts increase water input by increasing infiltration and soil moisture and reduce water output by reducing soil evaporation, thus eventually enhancing the water available to plants.

According to this observation, Bowker et al. [19] highlighted the importance of biological crusts development in arid landscapes, as they seem to be the most influential factor to reduce water erosion, and, also among the most manageable factor to achieve it through management practices.

2.3 Seed bank

It has often been seen that the introduction of herbivores in the Monte ecoregion has decreased the cover of grasses and their seed contribution, but very rarely has it been taken into account that these herbivores also reduced the cover of moss biocrusts, which would have provided suitable sites for seed germination, loading to the possibility of grassland regeneration. Some authors [39, 40] have already suggested that surfaces with biological crust constitute "safe sites" for the capture, germination, and establishment of seedlings, with greater humidity and more nutrients present, although it is not clear if all these processes are equally benefited. On the contrary, others (e.g. [38]) have observed that the cover of herbaceous species decreases with the increase in the abundance of biological crusts, and some even refer to them as weeds that compete with vascular plants [41].

In the Monte ecoregion, the dominant grasses are C3, and most are species of the tribe Stipeae. Two dominant perennial grasses of this tribe, *Nassella tenuis* (Phil.) Barkworth and *Piptochaetium napostaense* (Speg,) Hack, have a mechanism that allows their seeds to bury themselves in the ground during a rain, but, in order for the seeds to be able to be buried in that place, they must first be retained on the surface, awaiting the appropriate conditions. The surface roughness and the higher humidity of the biological crusts would provide the necessary conditions to retain these seeds and allow their anchorage.

To evaluate the effect of moss on the soil seed bank, we extracted soil samples with and without moss biocrust on the surface from the field with a hole puncher, and we removed the moss and the litter from them, then collected all the seeds of *Nassella tenuis* and *Piptochaetium napostaense* present in the samples by sieving and manual separation. The number of seeds buried in the first few centimeters of the soil was significantly higher for the samples with moss biocrusts than without, for both species evaluated (**Figure 8**).

In another field trial, we tested whether the moss biocrust acts as a seed trap, by laying a fixed number of intact propagules (seeds with sharp tips and twisted hydroactive awns) of *Nassella tenuis* on the surface of buried plastic rings, either with or without moss biocrust, after having manually removed all previous propagules [17]. We counted the seeds again 3 weeks later and found three times as many seeds in the samples with moss biocrust than without (**Figure 9**).

We repeated the field experiment of seed retention under greenhouse conditions supplying irrigation but with natural ventilation so that the seeds could potentially be moved by air currents inside it. Although this test was repeated twice, in spring and autumn, both times the samples with biological crust retained three times more seeds on the surface of the soil than the uncrusted plots ($p < 0.001$) (**Figure 10**), as had occurred in the field trial.

Our data contrast with those of Li et al. [42], who concluded that vascular plant seeds are not retained on the smooth moss-crust surfaces in windy environments, although our natural environments are very windy and, despite this, we found that the seeds were retained effectively, not only in the greenhouse trial. What remains to

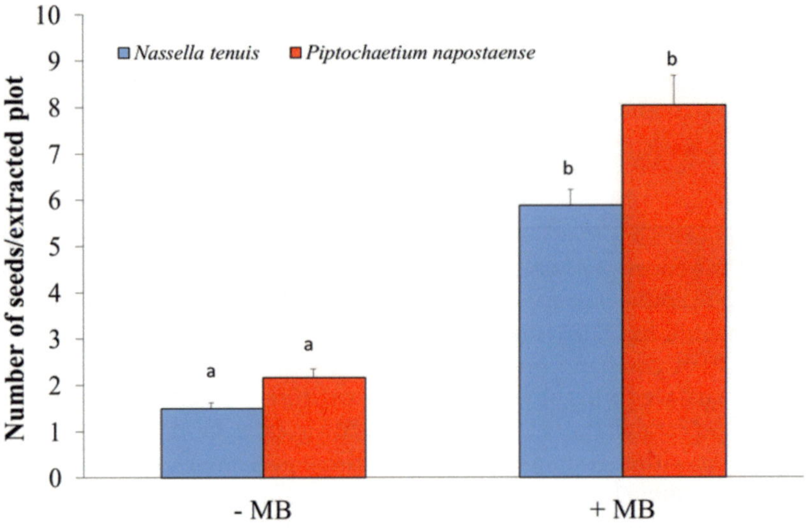

Figure 8.
Number of seeds per plot + SE of two grass species found in plots of 156 cm³, with and without moss biocrust. Lowercase letters indicate significant differences (p < 0.0001).

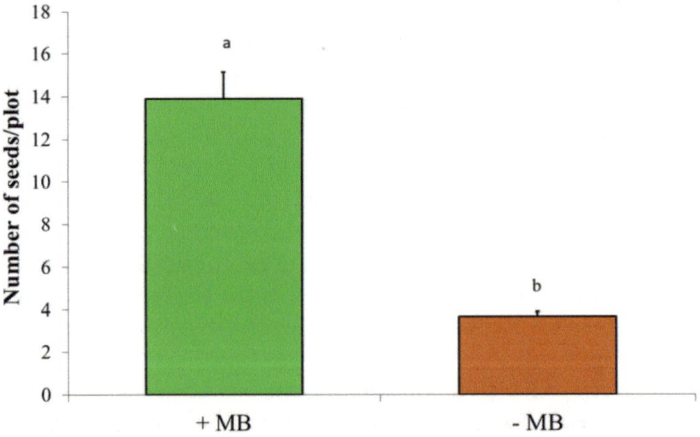

Figure 9.
Number of seeds of Nassella tenuis per plot of 45 cm² (+SE) with and without moss biocrust, retained after 3 weeks in the field. Lowercase letters indicate significant differences (p < 0.0001).

be tested is what happens with seeds of different sizes and shapes, since we have only analyzed what happens with dominant forage grass species.

2.4 Seedlings emergence, growth, and survival

If the seeds appear to be retained within the crust, what happens with their germination and the growth and survival of seedlings in that layer? There is no clear evidence about whether biological crust is a prerequisite for the development of higher plant cover or *vice versa*, or whether both processes occur simultaneously.

In a previous study [43], we found a greater number of seedlings in sites with moss biocrusts than without them, especially on flat surfaces between shrubs. This led us

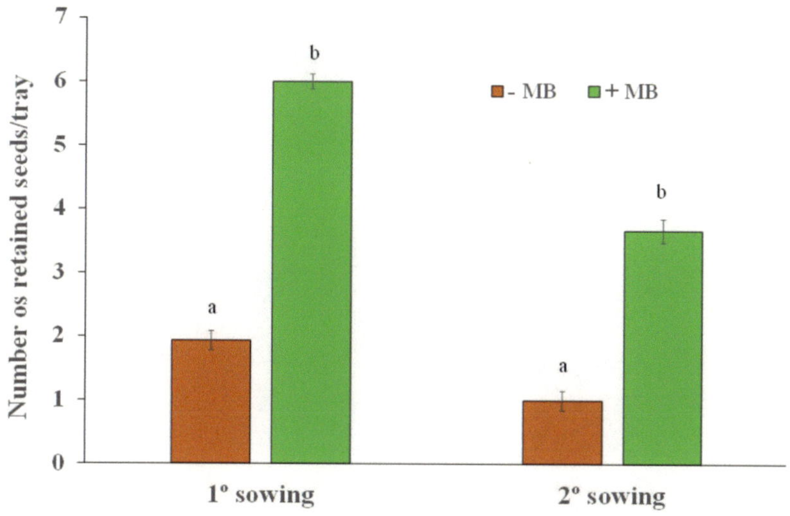

Figure 10.
Number of seeds ($\dot{x}$ ± SE) of Nassella tenuis retained on the surface of metallic trays of 375 cm^2 with (+MB) and without (−MB) moss biocrust, for two sowing dates. Lowercase letters indicate significant differences (p < 0.001).

to evaluate the survival of grasses in relation to the presence of moss biocrusts, so we carried out a trial in two different conditions, one in the field and the other under greenhouse conditions, with pots extracted from the field.

In the field trial, we marked *Nassella tenuis* seedlings within small plots with and without moss coverage and counted the number of tillers and leaves appearing during 8 months throughout the spring and summer. We found no significant differences in those variables between the two kinds of sites over that period, as we had expected ($p > 0.1$), given that it is known that the dark colors of many biological crusts would produce a more rapid rise in soil surface temperature in spring [44] which would contribute to accelerating germination, seedling growth, and the phenology in general of the vascular plants associated with them.

In the greenhouse trial, we watered the extracted pots up to field capacity until the seedlings germinated and then we interrupted irrigation; after a drying period, we watered the pots again and compared seedling survival. The emergence percentage was very low and did not differ between the two treatments ($p > 0.1$), but, growth, measured as number of tillers (**Figure 11**) and leaf length of seedlings (**Figure 12**), was significantly higher in pots with moss biocrust. Furthermore, 78% of the seedlings with moss produced new tillers *vs*. 65% of those without moss. In addition, seedlings leaf growth rate was significantly higher ($p < 0.0001$) in plots with moss biocrust [17], although seedling survival was similar in both treatments ($p > 0.1$).

When we analyzed the relative growth rate (RGR) in relation to leaf length, we found very strong relationships between them ($p < 0.0001$), but, although the negative trends were similar for both treatments, the seedlings growing in pots with moss biocrusts only reduced their RGR with longer leaf lengths than those in the samples without moss (**Figure 13**).

These results indicate that sites with biological crusts have surface water storage conditions that would favor germination by providing greater water availability and lower evaporation rates in the first few centimeters of soil. Once the plants were

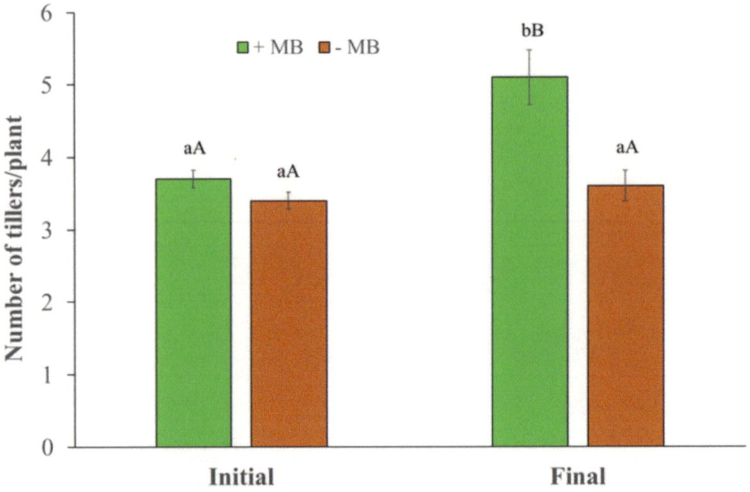

Figure 11.
Number of tillers (± SE), initial and final for plots with (+ MB) and without (– MB) moss biocrust. Lowercase letters indicate significant differences ($p < 0.05$) within the date and capital letters indicate significant differences ($p < 0.02$) between dates.

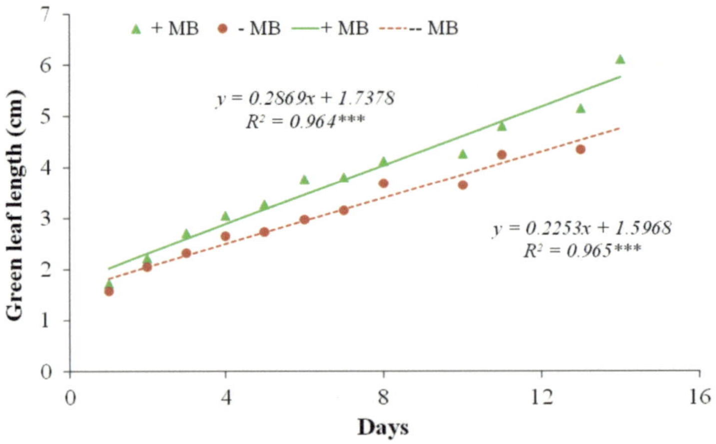

Figure 12.
Green leaf length of Nassella tenuis seedlings growing in pots with and without biological crusts in a greenhouse trial. Regression lines were highly significant ($p < 0.001$) and differ significantly between them ($p < 0.005$).

established, the pot trial suggested that plants growing on sites with biological crusts might have advantages in terms of increased growth rate; however, seedling survival did not differ between the two types of samples.

Trying to analyze plants survival in the field, we transplanted small plots with moss biocrust with and without seedlings of perennial grasses. We only watered them initially, at the time transplantation, then we left the pots subject to natural climatic conditions. A particularly dry period began after transplanting, so, a month after transplanting, we found that, although moss biocrust cover had prospered more in shady than in sunny sites (**Figure 14**), the patches generated by transplanting did not prosper homogeneously so as to allow measurements of seedling survival.

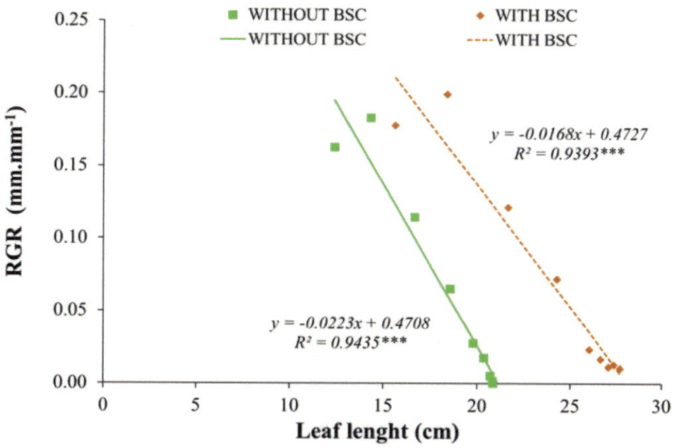

Figure 13.
Relative growth rate of Nassella tenuis seedlings with (+ MB) and without (– MB) moss biocrust in relation to leaf length. Regression lines were highly significant ($p < 0.001$). Source: Gustavo Cecchi.

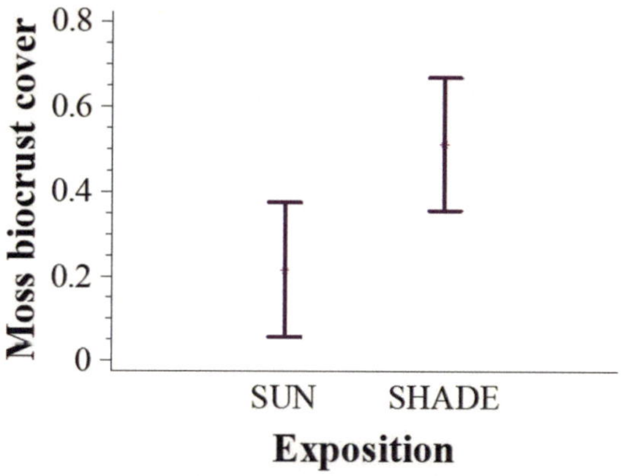

Figure 14.
Moss biocrust cover (arcsin percentage/100) of transplanted patches, in sunny and shady microsites of the field.

In general, we could only observe though not quantify that, as the survival of grasses and mosses was greater in the sites shaded by the shrubs than in the bare areas between them, the cover of moss biocrusts and the seedling density of the different species was not modified in the areas surrounding the transplanted patches.

This allowed us to visualize that, while biocrusts can be readily propagated under environmentally controlled conditions, rehabilitation in the field is complicated by environmental stresses which may be particularly acute in degraded, destabilized soils with harsh climatic conditions at the soil surface. However, it is a path that should be studied further, and certainly taking account of the most favorable period for transplanting in terms of weather conditions, and observing if these patches increase in soil coverage. In fact, today there are numerous efforts, by researchers and producers associated with them, trying to restore the crusts in drylands [45], and there is even a manual that synthesizes current information about biological soil crust restoration for resource managers making decisions on the ground (https://anitaantoninka.wixsite.com/biocrustrestoration).

3. Conclusions

In the eastern Monte ecoregion of Rio Negro province, in accordance with the results found by different researchers in other arid and semiarid regions of the world, the presence of moss biocrusts shows important functions related to the sustainability of the system, and this underlines the need to preserve them and even increase their coverage, in a way to contribute to the maintenance of ecosystem functions and mitigate the risks of climate change.

We have seen that disturbances of different scales, from livestock trampling to the removal of the shrub layer that may occur due to the action of man or the occurrence of wildfires, can affect moss biocrust cover. In the latter case, the damage will depend on the intensity of the fire. The use of an effective way to achieve moss biocrusts for ecosystems restoration should probably be studied further.

Regarding the use of water, this kind of crusts would improve infiltration, and therefore, the availability of water for the seedlings that may emerge there, in addition to the benefit that humidity provides to biocrusts themselves by allowing them to rapidly photosynthesize at the slightest wetting. The soils of our region are mostly clayey, with heavy textures, which favors their water retention capacity (in fact, the producers of the region refer to them as "bearing" soils), and this can be increased by the greater infiltration and less evaporation provided by the moss biocrusts.

On the other hand, the precipitation regime in our system is mainly autumn-winter-spring, which is different from other arid or semiarid ecosystems worldwide in which the main precipitation takes place in summer as torrential ones. The more extended period and the lesser rainfall intensity can contribute to water retention by these biocrusts, and this, added to the predominant soil's types mentioned, was reflected in the prolongation of the benefit that they contributed in terms of the lower evaporation that we found in pots with moss biocrust already extracted. These characteristics are also important when considering the reduction of water erosion risks in the system.

The greater humidity that moss biocrusts can conserve would also influence their possibilities of retaining seeds, and offering a favorable first environment for the emergence of seedlings and their initial growth. Perhaps, this humidity is the factor which reduces the risk of the seeds being blown away by the action of strong winds. Although we found beneficial results regarding seed retention, perhaps the benefit depends on the size and shape of the seeds that can fall within the biocrusts, and it would be necessary to determine if, at some point, the facilitation provided by moss biocrusts could be transformed into competition.

Apart from the researches previously mentioned that need to be deepened, our "black holes," in terms of regional information, are related to the contribution of moss biocrusts to nitrogen fixation and carbon sequestration, which should constitute future lines of research.

But there is no doubt that moss biocrusts should be taken into account when considering management practices for these systems, and the idea that their presence is a symptom of the "health" of these grasslands should be incorporated by both the researchers and producers.

Acknowledgements

To V. Alejandro Deregibus, the director of my master's degree, for his teaching and support; to Roberto Distel, for having helped me, discussing ideas with me, to Natalia

Villasuso, for having always collaborated in the field work. Many of these trials were carried out with financial aid from CONICET (National Council of Scientific Research) and the National University of Comahue.

Note

We carried out many of these trials with my husband, Gustavo Cecchi, who passed away long before seeing them written and published, although some of the hypotheses belonged to him.

Author details

Alicia Kröpfl
Regional University Center of Atlantic Zone, National University of Comahue, Viedma, Argentina

*Address all correspondence to: akropfl@yahoo.com.ar

References

[1] Langhans TM, Storm C, Schwabe A. Biological soil crusts and their microenvironment: Impact on emergence, survival and establishment of seedlings. Flora: Morphology, Distribution, Functional Ecology of Plants. 2009;**204**:157-168

[2] Belnap J, Lange OL. Biological Soil Crusts: Structure, Function, and Management. Heidelberg: Springer-Verlag Berlin; 2003

[3] Belnap J, Weber B, Büdel B. Biological soil crusts as an organizing principle in drylands. In: Biological Soil Crusts as an Organizing Principle in Drylands. Cham: Springer; 2016. pp. 3-13. DOI: 10.1007/978-3-319-30214-0_1

[4] Deng S, Zhang D, Wang G, Zhou X, Ye C, Fu T, et al. Biological soil crust succession in deserts through a 59-year-long case study in China: How induced biological soil crust strategy accelerates desertification reversal from decades to years. Soil Biology and Biochemistry. 2020;**141**:107665

[5] West NE. Structure and function of microphytic soil crusts in wildland ecosystems of arid to semi-arid regions. Advances in Ecological Research. 1990;**20**:179-223

[6] Condon LA, Pyke DA. Fire and grazing influence site resistance to Bromus tectorum through their effects on shrub, bunchgrass and biocrust communities in the Great Basin (USA). Ecosystems. 2018;**21**:1416-1431. DOI: 10.1007/s10021-018-0230-8

[7] Maestre FT, Benito BM, Berdugo M, Concostrina-Zubiri L, Delgado-Baquerizo M, Eldridge DJ, et al. Biogeography of global drylands. The New Phytologist. 2021;**231**:540-558

[8] Rivera-Aguilar V, Godinez-Alvarez H, Manuell-Cacheux I, Rodríguez-Zaragoza S. Physical effects of biological soil crusts on seed germination of two desert plants under laboratory conditions. Journal of Arid Environments. 2005;**63**:344-352

[9] Chamizo S, Cantón Y, Rodríguez-Caballero E, Domingo F. Biocrusts positively affect the soil water balance in semiarid ecosystems. Ecohydrology. 2016;**9**:1208-1221. DOI: 10.1002/eco.1719

[10] Xiao B, Zhao YG, Shao MA. Characteristics and numeric simulation of soil evaporation in biological soil crusts. Journal of Arid Environments. 2010;**74**:121-130

[11] Belnap J. Factors influencing nitrogen fixation and nitrogen release in biological soil crusts. In: Biological Soil Crusts: Structure, Function, and Management. Berlin, Heidelberg: Springer; 2001. pp. 241-261

[12] Zhao Y, Xu M, Belnap J. Potential nitrogen fixation activity of different aged biological soil crusts from rehabilitated grasslands of the hilly Loess Plateau. China. Journal of Arid Environments. 2010;**74**:1186-1191

[13] Su YG, Zhao X, Li AX, Li XR, Huang G. Nitrogen fixation in biological soil crusts from the Tengger desert, northern China. European Journal of Soil Biology. 2011;**47**:182-187

[14] Yang H, Li X, Wang Z, Jia R, Liu L, Chen Y, et al. Carbon sequestration capacity of shifting sand dune after establishing new vegetation in the Tengger Desert, northern China. The Science of the Total Environment. 2011;**478**:1-11

[15] Pagliai M, Stoops G. Physical and biological surface crusts and seals. In: Interpretation of Micromorphological Features of Soils and Regoliths. Elsevier; 2010. pp. 419-440. DOI: 10.1016/B978-0-444-53156-8.00019-2

[16] Kröpfl AI, Deregibus VA, Cecchi GA. Un modelo de estados y transiciones para el Monte oriental rionegrino. Phyton. 2015;**84**:390-396

[17] Kröpfl AI, Distel RA, Cecchi GA, Villasuso NM. Functional role of moss biocrust in disturbed semiarid shrublands of north-eastern Patagonia. Phyton. 2022;**20**:905-917

[18] Bowker MA, Belnap J, Miller ME. Spatial modeling of biological soil crusts to support rangeland assessment and monitoring. Rangeland Ecology & Management. 2006;**59**:519-529. DOI: 10.2111/05-179R1.1

[19] Bowker MA, Miller ME, Belnap J, Sisk TD, Johnson NC. Prioritizing conservation effort through the use of biological soil crusts as ecosystem function indicators in an arid region. Conservation Biology. 2008;**22**:1533-1543

[20] Anderson DC, Harper KT, Rushforth SR. Recovery of cryptogamic soil crusts from grazing on Utah winter ranges. Rangeland Ecology & Management/Journal of Range Management Archives. 1982;**35**:355-359

[21] Song G, Li X, Hui R. Effect of biological soil crusts on seed germination and growth of an exotic and two native plant species in an arid ecosystem. PLoS One. 2017;**12**:e0185839

[22] Rutherford WA, Painter TH, Ferrenberg S, Belnap J, Okin GS, Flagg C, et al. Albedo feedbacks to future climate via climate change impacts on dryland biocrusts. Scientific Reports. 2017;**7**:1-9

[23] Maestre FT, Escolar C, de Guevara ML, Quero JL, Lázaro R, Delgado-Baquerizo M, et al. Changes in biocrust cover drive carbon cycle responses to climate change in drylands. Global Change Biology 2013; **19**:3835-3847

[24] Duran P, Mora ML, Matus F, Barra PJ, Jofré I, Kuzyakov Y, et al. Biological crusts to increase soil carbon sequestration: New challenges in a new environment. Biology. 2021;**10**:1190

[25] Zaady E, Eldridge DJ, Bowker MA. Effects of local-scale disturbance on biocrusts. In: Biological Soil Crusts: An Organizing Principle in Drylands. Cham: Springer; 2016. pp. 429-449

[26] Concostrina-Zubiri L, Huber-Sannwald E, Martínez I, Flores Flores JL, Reyes-Agüero JA, Escudero A, et al. Biological soil crusts across disturbance-recovery scenarios: Effect of grazing regime on community dynamics. Ecological Applications. 2014;**24**: 1863-1877. DOI: 10.1890/13-1416.1

[27] Rostagno CM. Infiltration and sediment production as affected by soil surface conditions in a shrubland of Patagonia, Argentina. Journal of Range Management. 1989;**42**:382-385

[28] Kröpfl AI, Deregibus VA, Cecchi GA. Disturbios en una estepa arbustiva del Monte: cambios en la vegetación. Ecología Austral. 2007;**17**:257-268

[29] Zhang Y, Aradottir AL, Serpe M, Boeken B. Interactions of biological soil crusts with vascular plants. In: Biological Soil Crusts: An Organizing Principle in Drylands. Cham: Springer; 2016, 2016. pp. 385-406

[30] Bran DE, Cecchi GA, Gaitán JJ, Ayesa JA, López CR. Efecto de la severidad de quemado sobre la regeneración de la vegetación en el Monte Austral. Ecología Austral. 2007;**17**:123-131

[31] Brianne P, Rebecca H, David L. The fate of biological soil crusts after fire: A meta-analysis. Global Ecology and Conservation. 2020;**24**:e01380

[32] Muñoz-Rojas M, Delgado-Baquerizo M, Lucas-Borja ME. La biodiversidad y el carbono orgánico del suelo son esenciales para revertir la desertificación. Ecosistemas: Revista Cietifica y Tecnica de Ecologia y Medio Ambiente. 2021;**30**:2238-2238

[33] García-Carmona M, Lepinay C, García-Orenes F, Baldrian P, Arcenegui V, Cajthaml T, Mataix-Solera J. Moss Biocrust after Fire: Key for Soil and Microbial Recovery in Semi-Arid Mediterranean Forests? Available from: SSRN 4115347

[34] Noy-Meir I. Desert ecosystems: Environment and producers. Annual Review of Ecology and Systematics. 1973:25-51. California: Annual Reviews. Available from: http://www.jstor.org/page/info/about/policies/terms.jsp

[35] Belnap J. The potential roles of biological soil crusts in dryland hydrologic cycles. Hydrological Processes. 2006;**20**:3159-3178. DOI: 10.1002/hyp.6325

[36] Daryanto S, Eldridge DJ, Throop HL. Managing semi-arid woodlands for carbon storage: Grazing and shrub effects on above-and belowground carbon. Agriculture, Ecosystems and Environment. 2013;**169**:1-11. DOI: 10.1016/j.agee.2013.02.001

[37] Faist AM, Herrick JE, Belnap J, Zee JWV, Barger NN. Biological soil crust and disturbance controls on surface hydrology in a semi-arid ecosystem. Ecosphere. 2017;**8**:3. DOI: 10.1002/ecs2.1691

[38] Warren SD. Ecological role of microphytic soil crusts in arid ecosystems. In: Microbial diversity and ecosystem function: Proceedings of the IUBS/IUMS Workshop held at Egham, UK. 1993. pp. 199-209

[39] St Clair LL, Webb BL, Johansen JR, Nebeker GT. Cryptogamic soil crusts: Enhancement of seedling establishment in disturbed and undisturbed areas. Reclamation and Revegetation Research. 1984:238-243

[40] Eckert RE, Peterson FF, Meurisse MS, Stephens JL. Effects of soil-surface morphology on emergence and survival of seedlings in big sagebrush communities. Rangeland Ecology & Management/Journal of Range Management Archives. 1986;**39**:414-420

[41] Savory A. Holistic Resource Management. Island Press; 1988. p. 564

[42] Li FR, Kang LF, Zhang H, Zhao LY, Shirato Y, Taniyama I. Changes in intensity of wind erosion at different stages of degradation development in grasslands of Inner Mongolia. China. Journal of Arid Environments. 2005;**62**:567-585

[43] Kröpfl A, Cecchi G. Key sites for seedling emergence in degraded areas of the monte ecoregion of the Rio Negro Province, Argentina. MOJ Ecology & Environmental Sciences. 2022;**7**(3):89-94. DOI: 10.15406/mojes.2022.07.00252

[44] Xiao B, Wang H, Fan J, Fischer T, Veste M. Biological soil crusts decrease soil temperature in summer and increase

soil temperature in winter in semiarid environment. Ecological Engineering. 2013;**58**:52-56. DOI: 10.1016/j.ecoleng.2013.06.009

[45] Antoninka A, Faist A, Rodriguez-Caballero E, Young KE, Chaudhary VB, Condon LA, et al. Biological soil crusts in ecological restoration: Emerging research and perspectives. Restoration Ecology. 2020;**28**:S3-S8

Section 4

Bryophyta Biochemistry

Chapter 6

Protein Glycosylation in Bryophytes Differs Subtly from That in Vascular Plants

David Stenitzer and Friedrich Altmann

Abstract

Glycosylation substantially contributes to the physicochemical properties of proteins, and hence also cell walls. Moreover, they are key factors for the recognition of free or cell-bound glycoproteins by internal and external interaction partners. Green plants get by with a highly conserved, limited number of modifications of the pan-eukaryotic basic N-glycan structure. In detail, these are fucosylation of the innermost N-acetylglucosamine residue in 3-position, which renders plant glycoproteins immunogenic to mammals; xylosylation of the branching mannose; frequent occurrence of small N-glycans terminating with mannose or decoration of the antennae with Lewis A determinants. Bryophytes share all these features, but some mosses additionally display two peculiarities not seen in vascular plants. Many mosses exhibit 2,6-di-O-methylated mannose on the 6-arm and some mosses contain modified Lewis A termini with an as yet unspecified methyl pentose. Neither the responsible enzymes nor the function of these novel glycan features is currently known. Targeted glycoengineering of the moss *Physcomitrella patens* (Hedw.) Bruch & Schimp can allow the production of biopharmaceutical glycoproteins that are difficult to express in more established systems.

Keywords: glycoprotein, N-glycan, methyl-mannose, methylation, biopharmaceuticals

1. Introduction

In addition to their outstanding ecological importance, bryophytes have found numerous economic uses. On particular species, even has found applications for the production of recombinant proteins [1–3]. This has instigated the exploration of the potential for post-translational modifications in mosses in general and that of glycosylation in *Physcomitrella patens* (Hedw.) Bruch & Schimp in particular. At first, this moss appeared as a perfectly normal plant, as will be detailed in the next chapter. Only recently was it found that mosses in general have the ability to equip their proteins with features that have not yet been described elsewhere. In the last chapter, the differences between mammalian and moss O-glycosylation will be portrayed.

2. The commonalities between mosses and vascular plants

A number of investigations have shown that *P. patens* behave just like a tracheal plant when it comes to protein N-glycosylation. Biosynthesis and structures of plant N-glycans have been reviewed in many previous papers [4–6] and hence we only give–for the sake of comprehensiveness–a short overview. The biosynthetic routes from the ubiquitous oligomannosidic (*aka* high-mannose) structures lead to the check-point Man5Gn after which it branches out in various ways by combining a, however, limited, set of biosynthetic steps.

a. Removal of α1,2-linked mannoses by α1,2 mannosidase I.

b. Addition of a GlcNAc residue by GlcNAc-transferase I.

c. Removal of the α1,3- and (outer) α1,6-linked mannoses via the Golgi-resident α-mannosidase II.

d. Optional: addition of a GlcNAc residue in β1,2-linkage to the (inner) α1,6-linked mannose.

e. Addition of fucose in an α1,3-linkage to Asn-bound N-acetylglucosamine (GlcNAc) residue.

f. Addition of xylose in β1,2-linkage to the β-mannosyl-residue, whereby steps e) and f) occur independently.

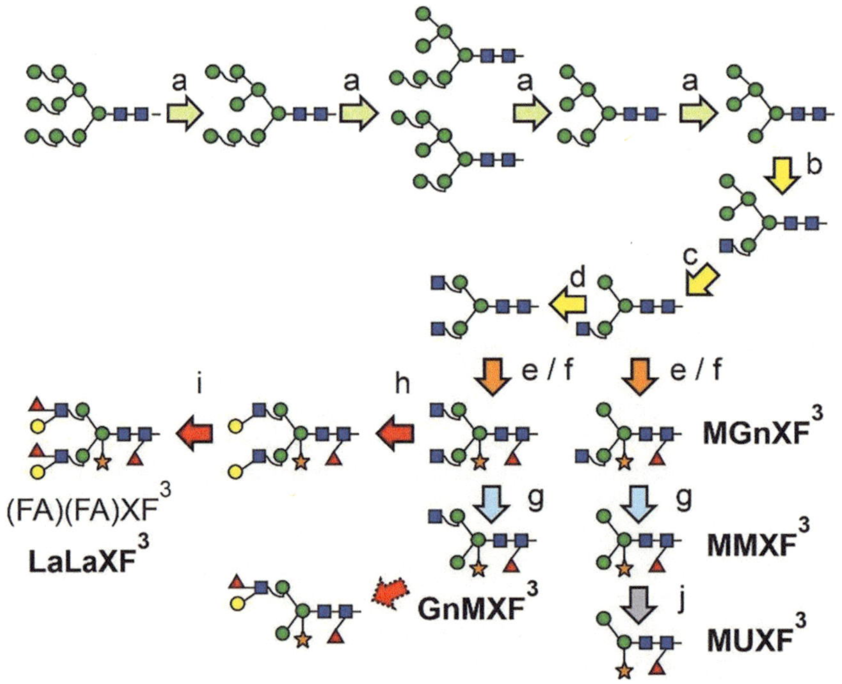

Figure 1.
Biosynthesis of typical plant-type N-glycans. The letters next to the arrows refer to the text above.

g. Removal of the GlcNAc linked to the α1,3-arm by hexosaminidase, for example, HEXO1 [7].

h. Addition of β1,3-linked galactose to any of the two GlcNAc residues, whereby this event is quickly followed by fucosylation.

i. Addition of β1,4-linked fucose to the sequence Galβ-3GlcNAc, thus forming Lewis A determinants also known as human blood group determinants.

j. Removal of the mannose linked α1,3 to the β-mannose. This occurs as a storage phenomenon in macerated plant material and generates the MUXF3 structure well known in allergy diagnosis [8].

A few main roads are depicted in **Figure 1**, which also gives names to the structures. To understand this naming system, we humbly ask the reader to remember that the "proglycan" nomenclature starts in the upper left corner and then lists the terminal residues in the counter-clockwise direction, whereby M, A, Gn, X, and F stand for mannose, galactose, *N*-acetylglucosamine, xylose, and fucose, respectively [9]. The branched Lewis A determinant of large plant N-glycans can either be written as "(FA)" or—more reader-friendly—as "La."

3. Methylation: a primordial resemblance

High-resolution mass spectrometry revealed the occurrence of small satellite peaks for MMXF^3 and MGnXF^3/GnMXF^3 in mass spectra of *P. patens* N-glycans, primarily when N-glycans were extracted from whole tissue (**Figure 2**) [10]. The mass increment of about 14 Da could also arise from the oxidation of a sugar moiety to glucuronic acid. The exact mass increment, however, and subsequent chemical analysis (linkage analysis via gas chromatography-mass spectrometry) clearly identified it as O-methylation of a mannose residue—more exactly, as 2,6-di-O-methylation. Collision-induced fragmentation in negative mode furthermore established the α1,6 linked mannose as being methylated (**Figure 2**).

Methylation in various ways was encountered in a previous study of *Chlorella*-clade microalgae [11–14]. Although methyl-mannoses were also found in these algae, none of the well over a dozen glycol-types contained complex-type/paucimannosidic structures and none of them contained 2,6-O-methylated mannose. As a recent unpublished survey of several *Scenedesmaceae* strains revealed likewise varied glycosylation (Mocsai R and Altmann F, unpublished results). We, thus, cannot exclude that other algae strains share with mosses this type of methylation and may thus be considered ancestors of mosses—at least glycosylation-wise.

4. Greater than following generations

The recent survey of N-glycosylation in a number of mosses surfaced several species that harbored N-glycans even larger than the fully developed Lewis A containing pride of the plant kingdom. The masses of the novel peaks indicated elongation of Lewis A determinants by 160 Da structures (**Figure 3**). Tentatively, we assume that a pentose plus methylation accounts for this mass increase. More precise information is not available so far.

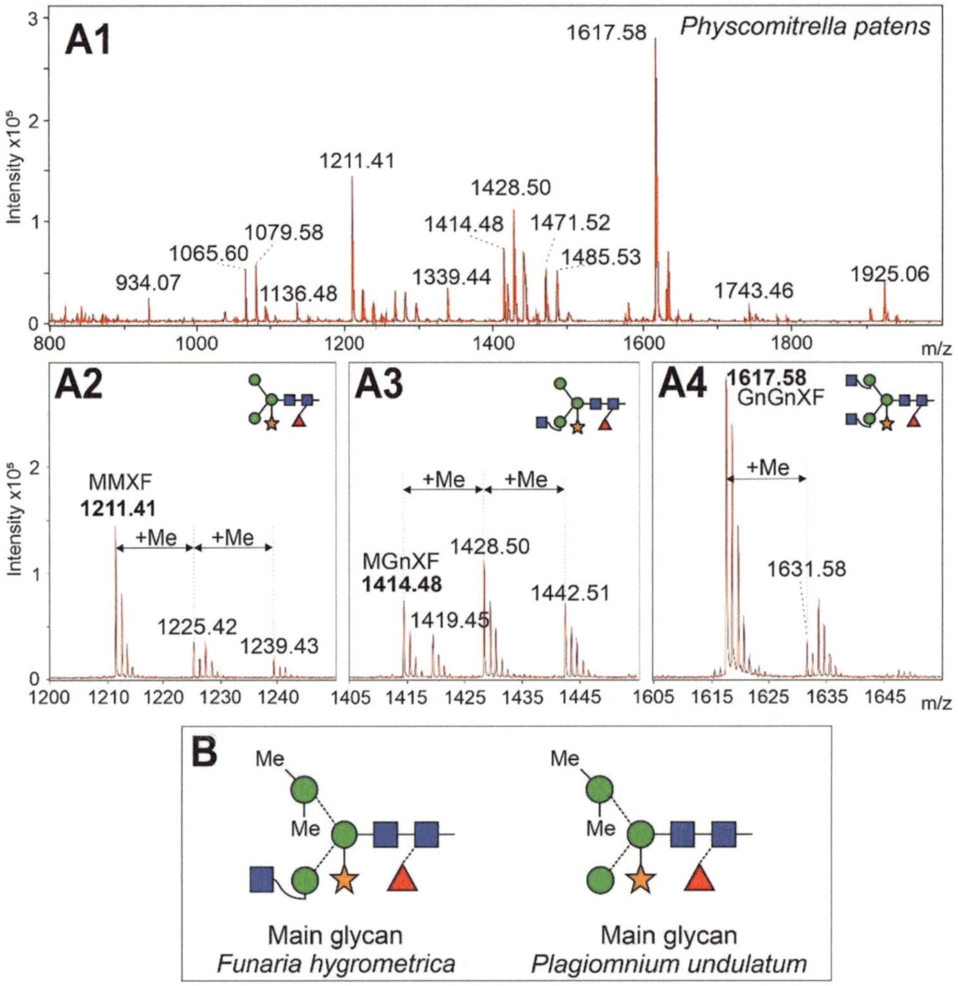

Figure 2.
Detection and structures of methylated N-glycans in mosses. Panels A shows the matrix assisted-laser desorption/ time-of-flight mass spectrometry (MALDI-TOF MS) spectrum of the N-glycans of P. patens in full width and in detail for the major affected N-glycan species. Panel B depicts the structures of the two examples that were structurally analyzed in detail derived from Funaria hygrometrica Hedw. and Plagiomnium undulatum (Hedw.) Kop.

5. Humanization of moss glycosylation and an odd interspecies confusion

All plants, including mosses, such as *P. patens*, are surprisingly apathetic towards switching off core fucosylation and xylosylation. Briefly, after the first cloning of the relevant enzymes fucosyl transferase and xylosyl transferase [15, 16], knock-down *Arabidopsis* and *Nicotiana* plants were generated [17, 18] and knock-out lines of *P. patens* were established for the production of biopharmaceuticals [19, 20]. The mainly resulting paucimannosidic structures, that is, N-glycans with two to five (a few = *pauci* in contrast to many = *oligo*) mannose residues and possibly one GlcNAc, xylose, and/or fucose residue were considered suboptimal for most purposes, and hence, efforts were undertaken to increase the levels of GnGn and to

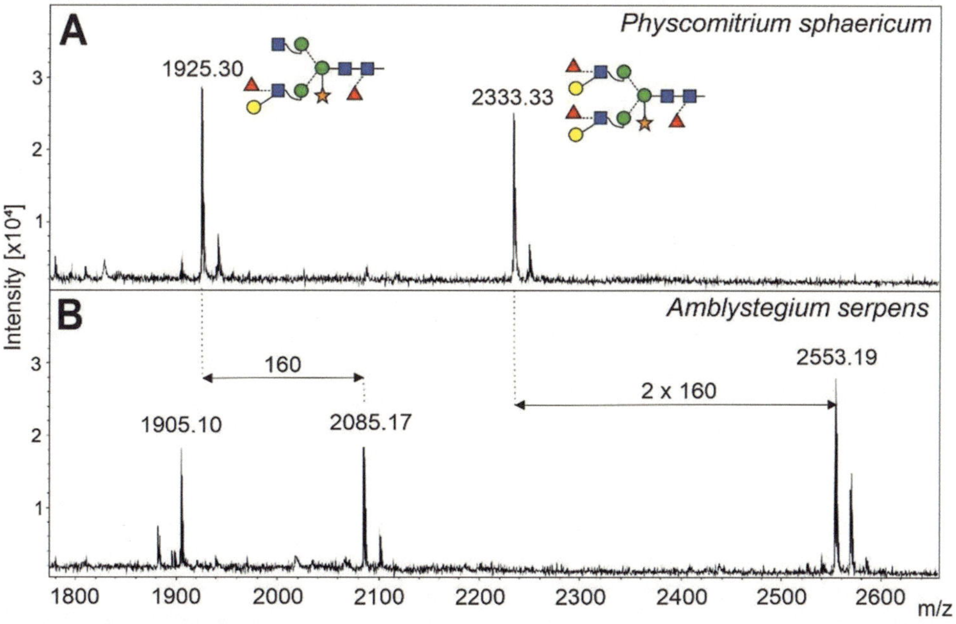

Figure 3.
Sections of mass spectra of two mosses representative of mosses with plant-typical Lewis A antennae (panel A) and those with larger glycans (panel B). Mass increments indicate the addition of a 160.07 Da element, which could be explained as the addition of a pentose and a methyl group. Panel A shows the high mass section of N-glycans from Physcomitrium sphaericum (Ludw) Fuernr and panel B that of Amblystegium serpens (Hedw.) Schimp.

eventually introduce the mammalian type core α1,6-fucose [21]. A big step toward humanized N-glycans was the introduction of β1,4-galactosylation [22]. The most stunning success, however, was the introduction of the entire mammalian sialylation pathway into plants—no less than six mammalian foreign genes that have to be actively expressed and correctly localized without harshly compromising the physiology of the host plant. This was at first realized with *Nicotiana* [23, 24] and then later with *P. patens* [3, 25]. As planned, these cloning steps led to the biosynthesis of substantial percentages of sialylated N-glycans. This established *P. patens* as in fact being very "patient", that is, tolerant and enduring towards this enormous intervention into the cellular machinery. Notably, certain otherwise difficult-to-express glycoproteins can be produced in suitable quality and quantity by glycoengineered *P. patens* [26].

However, next to some incomplete intermediate products, peaks with hitherto unknown mass levels occurred in mass spectra of moss lines expressing human β1,4-galactosyl transferase [27]. The mass increments of 132 Da indicated the addition of—against all rules—pentose. Being sensitive to α-arabinofuranosidase, this pentose was identified as furanosidic L-arabinose in α-linkage [27]. Its exact location was not known at the time of writing this chapter.

6. A mosses idea on O-glycosylation

The most often encountered type of protein O-glycosylation in mammals is the so-called mucin-type O-glycosylation, where Ser or Thr residues are at first decorated with *N*-acetylgalactosamine (GalNAc). Usually, this priming event is followed by the addition of other sugars, such as galactose, sialic acid, or GlcNAc [28]. In all plants, including mosses, this type of glycosylation is totally unknown. Instead, arabinans (chains of arabinoses) and arabinogalactans (AGPs, complex structures starting with a galactan that is substituted by arabinose chains and maybe other subtleties are found [29]. Arabinogalactans mostly occur as type II arabinogalactans [30, 31], but different architectures may also occur [32, 33]. Bryophytes do generate arabinogalactans, but with certain differences as compared to seed plant AGPs [34–36].

Notably, these "exotic" oligosaccharides are not linked to the codogenic amino acids Ser or Thr but to 4-hydroxyproline (Hyp) [29, 37]. Neighboring amino acids, in particular proline and hydroxyproline themselves, dictate if a given Hyp residue rather falls prey to galactosyl- or arabinosyl-transferase [29].

So, the initial step of O-glycosylation in plants is the oxidation of proline to hydroxyproline. The remarkable fact now is that apparently, the sites of mucin-type O-glycosylation of mammalian proteins are also the sites prone to be modified by prolyl-4-hydroxylase and then by arabinosyl-transferase as exemplified by human erythropoietin expressed in the moss *P. patens* [38] or human IgA1 [39]. Hardly surprising, the same holds true for *N. benthamiana* with the only, albeit technologically relevant difference, that several prolyl-4-hydroxylase are redundantly at work in vascular plants [39, 40], whereas knock-out of just one paralogous gene sufficed to suppress erythropoietin oxidation in *P. patens* [41].

7. Conclusion

The few bryophyte species whose protein glycosylation has been analyzed to date already presented some surprises. While first results indicated mosses to perform as their vascular relatives with regard to N-glycan biosynthesis, recent insights revealed them to present some peculiarities. Particularly interesting are mannose methylation and hyper-elongation of Lewis antennae.

Acknowledgements

This work was conducted at the University of Natural Resources and Life Sciences in Vienna, Austria. It was supported by the Austrian Science Fund (Doctoral Program BioToP; Molecular Technology of Proteins (W1224).

Conflict of interest

The authors declare no conflict of interest.

Author details

David Stenitzer and Friedrich Altmann*
Department of Chemistry, University of Natural Resources and Life Sciences, Vienna, Austria

*Address all correspondence to: friedrich.altmann@boku.ac.at

References

[1] Decker EL, Reski R. Current achievements in the production of complex biopharmaceuticals with moss bioreactors. Bioprocess and Biosystems Engineering. 2008;**31**(1):3-9. DOI: 10.1007/s00449-007-0151-y

[2] Decker EL, Reski R. Mosses in biotechnology. Current Opinion in Biotechnology. 2020;**61**:21-27. DOI: 10.1016/j.copbio.2019.09.021

[3] Decker EL, Reski R. Glycoprotein production in moss bioreactors. Plant Cell Reports. 2012;**31**(3):453-460. DOI: 10.1007/s00299-011-1152-5

[4] Faye L, Boulaflous A, Benchabane M, Gomord V, Michaud D. Protein modifications in the plant secretory pathway: Current status and practical implications in molecular pharming. Vaccine. 2005;**23**(15):1770-1778. DOI: 10.1016/j.vaccine.2004.11.003

[5] Strasser R. Plant protein glycosylation. Glycobiology. 2016;**26**(9):926-939. DOI: 10.1093/glycob/cww023

[6] Toustou C, Walet-Balieu ML, Kiefer-Meyer MC, Houdou M, Lerouge P, Foulquier F, et al. Towards understanding the extensive diversity of protein N-glycan structures in eukaryotes. Biological Reviews of the Cambridge Philosophical Society. 2022;**97**(2):732-748. DOI: 10.1111/brv.12820

[7] Liebminger E, Veit C, Pabst M, Batoux M, Zipfel C, Altmann F, et al. Beta-N-acetylhexosaminidases HEXO1 and HEXO3 are responsible for the formation of paucimannosidic N-glycans in Arabidopsis thaliana. The Journal of Biological Chemistry. 2011;**286**(12):10793-10802. DOI: 10.1074/jbc.M110.178020

[8] Altmann F. The role of protein glycosylation in allergy. International Archives of Allergy and Immunology. 2007;**142**(2):99-115. DOI: 10.1159/000096114

[9] Helm J, Grunwald-Gruber C, Thader A, Urteil J, Fuhrer J, Stenitzer D, et al. Bisecting Lewis X in Hybrid-Type N-glycans of human brain revealed by deep structural glycomics. Analytical Chemistry. 2021;**93**(45):15175-15182. DOI: 10.1021/acs.analchem.1c03793

[10] Stenitzer D, Mocsai R, Zechmeister H, Reski R, Decker EL, Altmann F. O-methylated N-glycans distinguish mosses from vascular plants. Biomolecules. 2022;**12**:136. DOI: 10.3390/biom12010136

[11] Mocsai R, Blaukopf M, Svehla E, Kosma P, Altmann F. The N-glycans of Chlorella sorokiniana and a related strain contain arabinose but have strikingly different structures. Glycobiology. 2020;**30**(8):663-676. DOI: 10.1093/glycob/cwaa012

[12] Mocsai R, Figl R, Sutzl L, Fluch S, Altmann F. A first view on the unsuspected intragenus diversity of N-glycans in Chlorella microalgae. The Plant Journal. 2020;**103**(1):184-196. DOI: 10.1111/tpj.14718

[13] Mocsai R, Figl R, Troschl C, Strasser R, Svehla E, Windwarder M, et al. N-glycans of the microalga *Chlorella vulgaris* are of the oligomannosidic type but highly methylated. Scientific Reports. 2019;**9**(1):331. DOI: 10.1038/s41598-018-36884-1

[14] Mocsai R, Kaehlig H, Blaukopf M, Stadlmann J, Kosma P, Altmann F. The structural difference of isobaric

N-glycans of two microalgae samples reveals taxonomic distance. Frontiers in Plant Science. 2021;**12**:643249. DOI: 10.3389/fpls.2021.643249

[15] Leiter H, Mucha J, Staudacher E, Grimm R, Glossl J, Altmann F. Purification, cDNA cloning, and expression of GDP-L-Fuc:Asn-linked GlcNAc alpha1,3-fucosyltransferase from mung beans. The Journal of Biological Chemistry. 1999;**274**(31):21830-21839. DOI: 10.1074/jbc.274.31.21830

[16] Strasser R, Mucha J, Mach L, Altmann F, Wilson IB, Glossl J, et al. Molecular cloning and functional expression of beta1, 2-xylosyltransferase cDNA from Arabidopsis thaliana. FEBS Letters. 2000;**472**(1):105-108. DOI: 10.1016/s0014-5793(00)01443-5

[17] Strasser R, Altmann F, Mach L, Glossl J, Steinkellner H. Generation of Arabidopsis thaliana plants with complex N-glycans lacking beta1,2-linked xylose and core alpha1,3-linked fucose. FEBS Letters. 2004;**561**(1-3):132-136. DOI: 10.1016/S0014-5793(04)00150-4

[18] Schahs M, Strasser R, Stadlmann J, Kunert R, Rademacher T, Steinkellner H. Production of a monoclonal antibody in plants with a humanized N-glycosylation pattern. Plant Biotechnology Journal. 2007;**5**(5):657-663. DOI: 10.1111/j.1467-7652.2007.00273.x

[19] Koprivova A, Stemmer C, Altmann F, Hoffmann A, Kopriva S, Gorr G, et al. Targeted knockouts of Physcomitrella lacking plant-specific immunogenic N-glycans. Plant Biotechnology Journal. 2004;**2**(6):517-523. DOI: 10.1111/j.1467-7652.2004.00100.x

[20] Parsons J, Altmann F, Arrenberg CK, Koprivova A, Beike AK, Stemmer C, et al. Moss-based production of asialo-erythropoietin devoid of Lewis A and other plant-typical carbohydrate determinants. Plant Biotechnology Journal. 2012;**10**(7):851-861. DOI: 10.1111/j.1467-7652.2012.00704.x

[21] Castilho A, Gruber C, Thader A, Oostenbrink C, Pechlaner M, Steinkellner H, et al. Processing of complex N-glycans in IgG Fc-region is affected by core fucosylation. MAbs. 2015;**7**(5):863-870. DOI: 10.1080/19420862.2015.1053683

[22] Strasser R, Castilho A, Stadlmann J, Kunert R, Quendler H, Gattinger P, et al. Improved virus neutralization by plant-produced anti-HIV antibodies with a homogeneous beta1,4-galactosylated N-glycan profile. The Journal of Biological Chemistry. 2009;**284**(31):20479-20485. DOI: 10.1074/jbc.M109.014126

[23] Castilho A, Neumann L, Gattinger P, Strasser R, Vorauer-Uhl K, Sterovsky T, et al. Generation of biologically active multi-sialylated recombinant human EPOFc in plants. PLoS One. 2013;**8**(1):e54836. DOI: 10.1371/journal.pone.0054836

[24] Castilho A, Strasser R, Stadlmann J, Grass J, Jez J, Gattinger P, et al. In planta protein sialylation through overexpression of the respective mammalian pathway. The Journal of Biological Chemistry. 2010;**285**(21):15923-15930. DOI: 10.1074/jbc.M109.088401

[25] Bohlender LL, Parsons J, Hoernstein SNW, Rempfer C, Ruiz-Molina N, Lorenz T, et al. Stable Protein Sialylation in Physcomitrella. Frontiers in Plant Science. 2020;**11**:610032. DOI: 10.3389/fpls.2020.610032

[26] Michelfelder S, Parsons J, Bohlender LL, Hoernstein SNW,

Niederkruger H, Busch A, et al. Moss-produced, glycosylation-optimized Human Factor H for therapeutic application in complement disorders. Journal of American Society Nephrology. 2017;**28**(5):1462-1474. DOI: 10.1681/ASN.2015070745

[27] Bohlender LL, Parsons J, Hoernstein SNW, Bangert N, Rodriguez-Jahnke F, Reski R, et al. Unexpected arabinosylation after humanization of plant protein N-glycosylation. Frontiers in Bioengineering and Biotechnology. 2022;**10**:838365. DOI: 10.3389/fbioe.2022.838365

[28] Jensen PH, Kolarich D, Packer NH. Mucin-type O-glycosylation--putting the pieces together. The FEBS Journal. 2010;**277**(1):81-94. DOI: 10.1111/j.1742-4658.2009.07429.x

[29] Kieliszewski MJ. The latest hype on Hyp-O-glycosylation codes. Phytochemistry. 2001;**57**(3):319-323. DOI: 10.1016/s0031-9422(01)00029-2

[30] Kaur D, Held MA, Smith MR, Showalter AM. Functional characterization of hydroxyproline-O-galactosyltransferases for Arabidopsis arabinogalactan-protein synthesis. BMC Plant Biology. 2021;**21**(1):590. DOI: 10.1186/s12870-021-03362-2

[31] Tan L, Varnai P, Lamport DT, Yuan C, Xu J, Qiu F, et al. Plant O-hydroxyproline arabinogalactans are composed of repeating trigalactosyl subunits with short bifurcated side chains. The Journal of Biological Chemistry. 2010;**285**(32):24575-24583. DOI: 10.1074/jbc.M109.100149

[32] Leonard R, Petersen BO, Himly M, Kaar W, Wopfner N, Kolarich D, et al. Two novel types of O-glycans on the mugwort pollen allergen Art v 1 and their role in antibody binding. The Journal of Biological Chemistry. 2005;**280**(9):7932-7940. DOI: 10.1074/jbc.M410407200

[33] Leonard R, Wopfner N, Pabst M, Stadlmann J, Petersen BO, Duus JO, et al. A new allergen from ragweed (Ambrosia artemisiifolia) with homology to art v 1 from mugwort. The Journal of Biological Chemistry. 2010;**285**(35):27192-27200. DOI: 10.1074/jbc.M110.127118

[34] Classen B, Baumann A, Utermoehlen J. Arabinogalactan-proteins in spore-producing land plants. Carbohydrate Polymers. 2019;**210**:215-224. DOI: 10.1016/j.carbpol.2019.01.077

[35] Fu H, Yadav MP, Nothnagel EA. Physcomitrella patens arabinogalactan proteins contain abundant terminal 3-O-methyl-L: -rhamnosyl residues not found in angiosperms. Planta. 2007;**226**(6):1511-1524. DOI: 10.1007/s00425-007-0587-y

[36] Bartels D, Baumann A, Maeder M, Geske T, Heise EM, von Schwartzenberg K, et al. Evolution of plant cell wall: Arabinogalactan-proteins from three moss genera show structural differences compared to seed plants. Carbohydrate Polymers. 2017;**163**:227-235. DOI: 10.1016/j.carbpol.2017.01.043

[37] Petersen BL, MacAlister CA, Ulvskov P. Plant protein O-arabinosylation. Frontiers in Plant Science. 2021;**12**:645219. DOI: 10.3389/fpls.2021.645219

[38] Weise A, Altmann F, Rodriguez-Franco M, Sjoberg ER, Baumer W, Launhardt H, et al. High-level expression of secreted complex glycosylated recombinant human erythropoietin in the Physcomitrella Delta-fuc-t Delta-xyl-t mutant. Plant Biotechnology Journal. 2007;**5**(3):389-401. DOI: 10.1111/j.1467-7652.2007.00248.x

[39] Mocsai R, Goritzer K, Stenitzer D, Maresch D, Strasser R, Altmann F. Prolyl hydroxylase paralogs in nicotiana benthamiana show high similarity with regard to substrate specificity. Frontiers in Plant Science. 2021;**12**:636597. DOI: 10.3389/fpls.2021.636597

[40] Uetz P, Melnik S, Grunwald-Gruber C, Strasser R, Stoger E. CRISPR/Cas9-mediated knockout of a prolyl-4-hydroxylase subfamily in Nicotiana benthamiana using DsRed2 for plant selection. Biotechnology Journal. 2022;**17**(7):e2100698. DOI: 10.1002/biot.202100698

[41] Parsons J, Altmann F, Graf M, Stadlmann J, Reski R, Decker EL. A gene responsible for prolyl-hydroxylation of moss-produced recombinant human erythropoietin. Scientific Reports. 2013;**3**:3019. DOI: 10.1038/srep03019